CAMBRIDGE LIBRARY COLLECTION

Books of enduring scholarly value

Technology

The focus of this series is engineering, broadly construed. It covers technological innovation from a range of periods and cultures, but centres on the technological achievements of the industrial era in the West, particularly in the nineteenth century, as understood by their contemporaries. Infrastructure is one major focus, covering the building of railways and canals, bridges and tunnels, land drainage, the laying of submarine cables, and the construction of docks and lighthouses. Other key topics include developments in industrial and manufacturing fields such as mining technology, the production of iron and steel, the use of steam power, and chemical processes such as photography and textile dyes.

Recollections of Forty Years

The French diplomat and engineer Ferdinand de Lesseps (1805–94) was instrumental in the successful completion of the Suez Canal, which reduced by 3000 miles the distance by sea between Bombay and London. This two-volume memoir, written towards the end of his life and dedicated to his children, was published in this English translation in 1887. In it, De Lesseps describes his experiences in Europe and North Africa. He includes reflections on European and colonial history and politics, an essay on steam power, and a report on the 1879 Paris conference that led to a controversial and abortive early attempt by a French company to build the Panama Canal. Volume 2 focuses on the Suez project, quoting extensively from De Lesseps' correspondence, and also contains facts and figures relating to the 'interoceanic canal', political essays, and the speeches for his inauguration into the Académie française.

Cambridge University Press has long been a pioneer in the reissuing of out-of-print titles from its own backlist, producing digital reprints of books that are still sought after by scholars and students but could not be reprinted economically using traditional technology. The Cambridge Library Collection extends this activity to a wider range of books which are still of importance to researchers and professionals, either for the source material they contain, or as landmarks in the history of their academic discipline.

Drawing from the world-renowned collections in the Cambridge University Library, and guided by the advice of experts in each subject area, Cambridge University Press is using state-of-the-art scanning machines in its own Printing House to capture the content of each book selected for inclusion. The files are processed to give a consistently clear, crisp image, and the books finished to the high quality standard for which the Press is recognised around the world. The latest print-on-demand technology ensures that the books will remain available indefinitely, and that orders for single or multiple copies can quickly be supplied.

The Cambridge Library Collection will bring back to life books of enduring scholarly value (including out-of-copyright works originally issued by other publishers) across a wide range of disciplines in the humanities and social sciences and in science and technology.

Recollections of
Forty Years

VOLUME 2

FERDINAND DE LESSEPS

CAMBRIDGE
UNIVERSITY PRESS

CAMBRIDGE UNIVERSITY PRESS

Cambridge, New York, Melbourne, Madrid, Cape Town, Singapore,
São Paolo, Delhi, Dubai, Tokyo, Mexico City

Published in the United States of America by Cambridge University Press, New York

www.cambridge.org
Information on this title: www.cambridge.org/9781108026406

© in this compilation Cambridge University Press 2011

This edition first published 1887
This digitally printed version 2011

ISBN 978-1-108-02640-6 Paperback

RECOLLECTIONS OF FORTY YEARS.

VOL. II.

RECOLLECTIONS OF

FORTY YEARS

BY

FERDINAND DE LESSEPS

TRANSLATED BY C. B. PITMAN

IN TWO VOLUMES

VOL. II.

LONDON: CHAPMAN AND HALL
LIMITED
1887

CONTENTS.
VOL. II.

———•———

vi   *CONTENTS.*

CHAPTER X.

PAGE

ABD-EL-KADER 236

CHAPTER XI.

ABYSSINIA 242

CHAPTER XII.

THE ORIGIN AND DUTIES OF CONSULS 273

CHAPTER XIII.

THE FRENCH ACADEMY 286

RECOLLECTIONS OF FORTY YEARS.

CHAPTER IV.—*Continued.*

THE ORIGIN OF THE SUEZ CANAL.

Journey to the Soudan.

I.

" AS soon as an International Commission of Engineers had fixed the mode of making the canal, and pointed out the preliminary works which should be undertaken before entering upon the enterprise itself, the British Government showed itself hostile to the project, and made overtures at Constantinople for a change in the order of succession, representing Mohammed Said as bereft of his senses. The Prince got wind of this, and confided to me how uneasy he felt. So, in order to escape the worrying of the English agents, he suggested that I should go with him to the Soudan. He was anxious to deliver that country from the misery and oppression by which it had been weighed down since the conquests and administration of Mehemet Ali. During our absence the investigations and preliminary works were to be continued in the desert of the isthmus, more than twenty leagues from any dwelling-place or travelling route, without regard to any protest from London or Constantinople.

" A flotilla of ten steamers was soon ready for his Highness, his ministers, his staff, two battalions of infantry, and a few guns. I was to start afterwards, and the Viceroy was to wait for me at Siout. My steamer was still moored to the quay at Boulak on the 26th of November. About midnight I had gone to my cabin on deck, and was just getting into bed when a candle set fire to the mosquito curtains and enveloped me in the flames. I endeavoured to put them out, as I could not open the door at first, owing to the bolt being so rusty, and failing in this, and fearing that I should be suffocated, I summoned all my strength and managed to burst open the door. I rushed on deck, ordered the captain to cut off all communication with the land, and to start at once. Part of my body was one large wound, and there were several lesser burns upon my legs. I was carried on to a bed, and there, after having had applications of tallow placed upon the flesh where the skin was gone, I made the attendant pour the beneficent Nile water over the sore places. Thanks to the care and company of my travelling companions, Dr. Abbate, physician to the family of the Viceroy, the French engineer, Motet Bey, and my secretary and interpreter,

Vernoni, I did not even have an attack of fever. But when, on reaching Siout, the Viceroy came to see me, I found it impossible to rise. I told him that my accident was of good omen for the rest of the journey, as we had acquitted our debt to ill-luck. We had a long and interesting conversation upon the results anticipated from our distant excursion. He was anxious to abolish slavery in the centre of Africa, and prepare in Ethiopia a trade which would be beneficial to the Suez Canal. He wished to appear as a sovereign benefactor in the region where his brother, Ismail Pasha, had been massacred with all his staff.

" It was forty years since Mehemet Ali, after having delivered Egypt from the oppression of the Mamelukes, had sent his second son Ismail to the Soudan, keeping his eldest son Ibrahim in Egypt to commence the formation of a regular army, with the aid of a French officer, Sèves, who, under the name of Soliman Pasha, became celebrated in the campaigns of Euboea, Morea, and Syria. Prince Ismail required at the outset of his campaign that a thousand slaves, a thousand camels, a thousand measures of wood, a thousand loads of hay, etc., should be brought to his camp.

" The inhabitants were obliged to submit, but while they brought him the tribute they were at the same time conspiring to rid themselves of him. One day, while he and his staff were enjoying a luxurious repast, the insurgent chiefs surrounded his camp with a belt of faggots, to which they set fire in the middle of the night, and the Egyptians who endeavoured to escape were massacred by the Soudanese.

"Vengeance for this was entrusted by Mehemet Ali to his son-in-law, the Defterdar, who committed atrocities the very description of which makes the blood boil. I am told that he was equally cruel to those of his soldiers or servitors who were lacking in discipline.

"Upon one occasion, at the request of a woman of the country, who came to complain that an Egyptian soldier had stolen some milk, he sent for the man whom she accused, having first warned her that he would have her ripped open if she had told a falsehood. The soldier was then ripped open, and as his stomach was found to contain traces of milk the woman was dismissed with a largess. Upon another occasion, as his horse was badly shod, he sent for his *saïs* (running groom) and had the horse's shoes nailed to his feet.

"The Defterdar scattered terror and desolation throughout the Soudan, leaving nothing but ruins behind him, and bringing back to Egypt a hundred thousand slaves. It is easy to imagine how miserable and oppressed were the populations which had remained since then beneath the military authority of the rapacious Turkish governors.

"Such is the country which Ismail's brother and

the brother-in-law of the Defterdar is about to take me through. When he left Siout with his suite it was arranged that I should rejoin him on the 18th December at Korosko, between the first and second cataract, but as my wounds were not entirely healed he went on in advance, and arranged to meet me at Berber, above the last cataracts of the Nile.

II.

"Upon December 24th I was still unfit to walk, but I got myself hoisted on to my dromedary, to cross in six days this same desert of Korosko. We had to guide us on our way the skeletons of the camels which had long since been abandoned by passing caravans. The entire bodies of the camels which had been left behind during the passage of the Viceroy, though quite dried up, still were in the same position as that in which they fell beneath their burden. Birds of prey were seen creeping out of their bodies, and jackals were patiently waiting in the distance until the vultures had done their meal to come and finish up the remains. We halted for half a day near a well in the middle of the desert. This point is the only one from which, at this season of the year, the four stars of the Southern Cross could be seen in the Southern hemisphere, and the North Star in our hemisphere. While waiting to observe these stars, which were not to be visible till between two and three in the morning, I amused myself by getting the Arab chiefs to tell some of their Eastern stories. One of these struck me very much, because of the very delicate sentiments which it expressed as to the superior morality of woman. Here it is very prosaically translated :—

" ' A moth was in love with the light. Incessantly attracted towards it, the moth flew close up to it. But no sooner had the tip of its wing been slightly scorched than it flew off again, throwing itself at the feet of the cruel one, filling the air with its plaintive cries.

" ' In the meanwhile the light was dying out; before throwing out its last flicker it said to its lover : " Moth, you have made much ado about a slight singeing of your wings ; you have re-proached me unjustly ; I have loved you in silence ; my flame is about to expire ; I am dying. Adieu. Fly to other loves ! " ' '

" Our caravan started again at an early hour, after having had the satisfaction of contemplating in all their splendour the Southern Cross upon the one side and upon the other the North Star, an old friend who had often guided me in my voyages through the desert. Having reached the banks of the Nile at Abu-Hamet, on January 1st, 1857, I was anxious to get to Berber before nightfall, in order to wish a happy new year to the Viceroy. I hurried my drome-dary forward, and did seventy-five miles in the day. I found the Prince alone in his tent, crying bitterly

I asked him what was the matter, and he said that his generals had just put the same question to him. ' I told them,' he went on to say, ' that the music had affected my nerves ; but I will confide to you that I am weeping over this unfortunate country, which my family has made so wretched ; and when I think that there is no remedy for all this it afflicts me sorely.' I endeavoured to console him by pointing out to him that, on the contrary, there were remedies which he, with his spirit of justice, would be able to discover and apply.

" The next day we started for Shendy, the very place where his brother Ismail had been burnt to death. The Viceroy had appointed this as the place where all who had presented petitions to him in the course of his journey were to meet ; and upwards of one hundred and fifty thousand natives were assembled there. In the presence of this vast multitude the prince was informed that, despite his formal injunctions, an aged Turkish chief had detained a female slave chained up in a cave. He gave orders for master and slave to be brought before him, had the chains transferred from the one to the other, and thus excited extraordinary enthusiasm. Carried away by the popular applause, he told the people to remove the cannon from the citadel and cast them into the Nile ; but on my whispering to him that perhaps this was trusting them too far, he said to me, ' The guns are too old ; they were placed there in my

father's time, and are incapable of firing a single
shot.'

" The Viceroy then declared that he intended to
send all the Turkish functionaries back to Egypt;
that he should leave them to govern themselves ; and
that he intended to establish among them municipalities, which had from the beginning of the world
been the principal element of all organised society.

" I was instructed to remain a few days at Shendy
to assist his Highness's Ministers in the creation of
the municipalities, which were formed by election
from among the heads of families.

" Boats were got ready to take us up to Khartoum,
where we arrived on the evening of January 10th. The
name Khartoum signifies the two branches of the elephant's trunk, because the town is situated between
two tusks, as it were—the Blue and the White Nile.
I am met on arrival by the Viceroy, who is waiting
for me at the entrance to the audience chamber in the
palace of the Governor-General of the Soudan.   He
tells me that he had been greeted, as mentioned by
me in a previous chapter, on his arrival by a band
of music such as he had never heard before, the
wind instruments in which, dating from the time of
Mehemet Ali, had been mended with soap plaster
borrowed from the regimental chemist.

" I embarked upon my voyage up the White Nile
with Arakel Bey, a very amiable and intelligent
young man, who had been brought up in France at the

Collège de Sorrèze,* and who was very ambitious to be of service to his country. As we went along the banks of the Blue Nile in order to enter the White Nile, we saw long files of dromedaries coming in from all directions, mounted by men of every shade of colour, from chocolate to ebony black, who had hurried to Khartoum from the most remote districts to thank the great prince whose fame had traversed the desert, and who came to bring freedom to the oppressed.

"In the first bark there were, in addition to Arakel Bey and myself, M. Heuglein, Austrian Consul at Khartoum, and a very learned explorer and naturalist, and Senhor Popotani, Consul-General of Portugal in Egypt, for whom the Viceroy had a great liking. In the second bark were some of our services, the provisions, and the cooking apparatus. We were becalmed all night at the junction of the two streams. The next morning a brisk wind took us up to about the 15th degree, to the south of Mount Oueli. The White Nile is at this point two or three times as broad as the river is in Egypt or Nubia. Its banks are not steep—that is to say, the river is not embedded between two high banks—and the ground covered with timber slopes gradually down to the edge of the stream. M. Heuglein tells us that the river, with its numerous islands, was much the same up to the fourth

---

* Note of the Translator.—This was the college founded by Père Lacordaire.

degree, which is at present the extreme limit known. We encountered flocks of waterfowl—the sacred ibises, which are no longer to be seen in Egypt, royal cranes, grey cranes, Nile geese, and pelicans. At about two o'clock, the wind having dropped, we let our barks drop down stream, and while they were running down, we landed on the right bank, about two leagues to the south of Mount Oueli. We made for the direction of the mountain, following some very densely-wooded paths, and Arakel Bey and myself went up the mountain while the two others were shooting game. From this height and in so clear an atmosphere we saw, for a distance of ten or twelve leagues all round us, plains covered with forests, and natural vegetation which could, with the facilities for irrigation, be made of enormous value.

"Upon coming down from the mountain, we all assembled at an encampment of the *Bindja* tribe. The sheik and his family received us very cordially, and the most elegant of the cocoa-nut mattings were taken down from the walls and placed at our feet. We were treated as personages belonging to the suite of the Viceroy, whose deeds of benevolence are already known throughout the country, and who is called the ' Father of the Unhappy.' The women—who, despite their colour of Florentine bronze, are very handsome —bring us milk and fruit. Old men, surrounded by their families, sing the praises of the Effendinah (our

master), and pray aloud for him, prostrating themselves on the ground, and exclaiming that God had
sent him to deliver them from their misery.

" At eight o'clock the barks came to fetch us at the
place where they saw the fires alight. While we
were having supper, as I happened to praise the taste
of the excellent Bindja milk, M. Heuglein made me
feel rather uncomfortable by telling me that upon
the Upper Nile the tribes which have no salt mix
the cows' urine with milk. He added, however, that
this custom only commenced with the tribes about a
hundred leagues higher up the river."

III.

"January 18, 1857.

" Upon the morning of the 16th we were still only
ten leagues from Khartoum. There was a very slow
current and no wind, so the boats went slowly up
stream. In the afternoon we landed and walked
through some woods and some bean fields in flower,
which emitted an odour which was very pleasant at
first, but soon became too strong. The geese, cranes,
and herons swept down upon the banks of the river,
and looked in the distance like flocks of sheep, but
they would not let us get within gun-shot of them.
I was walking on ahead, accompanied only by the
boatman, when, as we approached a small creek, we
noticed two sharp points floating on the surface of the
water and making for land. We saw, as we got
nearer, that these were the muzzles of two crocodiles

which were swimming about on the look-out for some prey. When about fifteen paces off I fired at one with a rifle, but the bullet sounded on the animal as on a piece of wood, and the beast did not move. My boatman told me that if any woman or child, or even a man alone, came to fetch water just then, he or she would incur a great risk of being seized. He added that when a crocodile attacks it begins by taking the victim under its claws and squeezing it tightly, dragging the body off to devour it upon some neighbouring island.

"He went on to tell me that, being one day in the water and swimming about with his brother, one of their comrades who was on shore called out to them to be careful, as he had just seen a crocodile. The two swimmers at once made for shore, but their comrade incautiously had advanced close to the edge of the water, and the crocodile, making a prodigious bound, seized him by the left arm, plunged into the stream and came up on the other side, where my boatman distinctly saw him devour the body of his unfortunate comrade. He also showed me a wound which a crocodile had made in his leg. He once met one which had gone ashore and was waddling back to the Nile. He and his companion tried to stop it, but the crocodile came at him, and with its open jaw inflicted a bite which threw him to the ground. Fortunately, he had the time to seize the dagger which the natives wear in the form of a bracelet, and with this he succeeded in wounding the crocodile in the vulnerable part of the neck, which has no scales, whereupon the animal made at once for the Nile. He told me that there was another way to make the crocodile let go of you if he seized you in the water, and that was to push your fingers into his eyes, if your position allowed you to do so.*

"We re-embark and continue our journey down the river, remaking several traces of the hippopotamus. It is evident that we are in the region frequented by these amphibious creatures, and we soon see in mid stream a sort of floating island, blackish in colour, and with its surface shining in the sun. This was the back of an enormous hippopotamus. We soon saw another one not so large. When we got quite close to the larger one, the sailors shouted in a peculiar manner, and we saw the hippopotamus rapidly plunge to the bottom, and then come up again to the surface and expose all the upper part of his body and the hind legs. We were told that this was a family party, and that the mother, believing her young to be in danger from the boats, had sprung out of the water in this way to see what her enemies were and, if necessary, defend herself.

"This reminded me of a story which had been told me, upon my arrival at Khartoum, by Father Knoble-

---

* Note of the Translator.—This must be almost as effective a mode of self-preservation as putting salt upon birds' tails is of catching them.

cher, superior of the Catholic Mission upon the White Nile. On one of his voyages the boat in which he was travelling having separated the mother from her young, she jumped furiously out of the water, and as Father Knoblecher's cook happened to be leaning over the side of the boat, he was struck by the enormous beast as she fell back and dragged him with her into the stream.

"We reached Khartoum at nine in the evening of the 17th, and the next day the Viceroy informed me that he had dictated during my absence his ordinances for the administration of the Soudan. These curious documents remind one at once of the ancient ordinances of the French kings, and the patriarchal traditions of the Bible. A few fragments of them are worth quoting:—

" ' *Order of His Highness the Viceroy to the new Governors of the five provinces of the Soudan: Sennaar, Kordofan, Taka, Berber, and Dongola.*

(Translated from the Arab.)

" Khartoum, *January* 26, 1857.

" ' You have heard what my heart yearns for, and how I desire the prosperity of the land and the welfare of the population. You know also how I have sought to form a right understanding of whatever is calculated to develop their fortune, to spare them suffering and place them beyond the reach of persecution, so that they may reach the height of prosperity by the removal of injustice and of the abuse of power.

" ' When I reached the Soudan provinces and saw the misery in which they were plunged, owing to the excessive sums levied upon the lands, I decided, moved by the spirit of justice, that all this system should be abandoned, and I desire that henceforth the taxes shall be distributed according to the means of the inhabitants, so that all fears may be calmed, that the land may prosper, and that there may be no further cause for complaint or exasperation.

" ' When I reached Berber I asked the sheiks and inhabitants who came out to meet me what could insure their tranquillity, and how much they could afford to pay. They replied by asking that each sakié should pay an import of 250 piastres; but as my love for my people makes me desirous of giving them the utmost possible prosperity, and as I am anxious to restore confidence to those who have expatriated themselves and induce them to return, I have decided that they shall pay only 200 piastres for each sakié. I then arrived at Khartoum to meet the other sheiks and notable persons, and if these latter had arrived promptly, they would have experienced, by the effect of my presence among them, the marks of a generosity which they had never yet experienced. But as I have made you Mudir of this province, you must above all things concern yourself with the welfare of the populations, with all that can ameliorate their position and tranquillise their minds ; and you are to act in regard to them with all possible solicitude.

" 'You will collect the taxes at the time of the most profitable crops—that is to say, that every year you will call together an assembly during the three months when there is no labour to be done in the fields. At this meeting you will divide the payment of the taxes into monthly sums, so arranged that they will not be burdensome to the inhabitants or leave arrears behind. This assembly is to be composed of from twelve to twenty-four notables of the province, according as you shall deem best for the general good. In your position as president of this assembly it will be your duty to see as to the division of the taxes, the best means for increasing the general welfare and tranquillity, so as to render the state of the towns and villages very stable. Your decisions are to be communicated to me from time to time. . . . .

" 'Whatever the Government may require in the way of food, camels, or labour is always to be paid for at the rate of two per cent. over what the inhabitants pay for the same things ; and even if it should happen that the value and the hire of the articles increased, the Government is always to pay the extra two per cent. ; and in order to guard against the sheiks, with the view of showing that they are watching over the interests of the Government, not declaring the truth for the price and hire of labour, you shall not take anything except with the free consent of the owners, so that in this way prosperity may increase, and that others, seeing the price paid by the Government, may be led themselves to pay more, which is the way to increase the welfare of the country. You will take no man or camels for corvées (forced labour); you will advise the inhabitants to sow wheat, indigo, cotton, and sesamum. You will do all that is necessary to see that the cottons are properly pressed and the indigo well made so as to facilitate their export and increase their value. You will also encourage the inhabitants to extract sesamum oil, for that is in their interests. There are also many forests which contain an immense quantity of wood suitable, some for building, some for boat-making, some for firewood. It would be easy to send this timber down to Egypt on a raft when the Nile rises. You must let the inhabitants understand this and encourage them to do it, for most of them have little to do, and this would be a fresh source of profit for them. . . . .

" ' With regard to the mountains which are taxed, as their inhabitants live like savages, and as it is necessary to bring them to a state of humanity, so that they may no longer be inclined to revolt, I have decided to forego two-thirds of their taxes. You will explain to them that they are not slaves, but free. These persons are in the habit of sowing some of the land on the slope of the mountains. You must encourage them and make them understand the advantages of life in towns; exhort them to increase their cultivation, and endeavour to convince and attract them. Explain well to them that if they heartily devote themselves to agriculture I will dispense from payment of the tax which I now reduce, and thus they will only have to pay the tax of the lands which they actually cultivate, even if this tax should be less than what they pay for their mountains; and you will treat them in this manner for their tranquillity, and so as to draw them into the path of civilisation. If even, in your conversation with them to explain this and to prevail upon them to do it, they ask you to remove this tax, provided that they promise to devote themselves to agriculture, paying only the land tax, you will consent and will refer the matter to me, so that I may act with them according to their desires, with the sole object of inspiring them with the love of comfort of life in the towns, and to safeguard them from the vicissitudes to which they are exposed.

"'When I arrived at Berber and at Shendy, I appointed the sheiks and notables according to the wishes of the inhabitants and at their choice. The sheiks of some villages did not come. You will arrange things in the same way for the province of Dongola, and complete them also for the villages in the provinces of Berber and Gaulein, where they have not been done. You will select as sheiks and muluks those who have been chosen by the inhabitants, and you will give them your wise counsels, so that they may behave properly and avoid, thanks to your care, anything which might alienate the inhabitants. Examine all affairs submitted to you; do justice to all men without partiality and in all equity. If any man deserves imprisonment for any misdeed, you will have the matter tried at once, so that the culprit may not remain long in prison; for even when it is necessary to punish a man for a bad action, so that he may not again fall into evil, my pity and clemency would not have him remain in prison longer than is absolutely necessary. Although, considering all that I have just done in favour of the inhabitants of this country, either by diminution of the taxes and the abolition of forced labour, or by preventing injustice and oppression, it does not seem necessary to maintain troops there, inasmuch as the inhabitants will necessarily be compelled, for the preservation of their properties, to defend themselves against attack, I have nevertheless quartered a sufficient number of regiments in the different localities. Be on your guard, therefore, to repel whomsoever attacks you, and if it is necessary that the provinces should come to one another's help, let this be done so that no harm may befall any of those under your charge.

"'It is always a matter of urgent necessity, and it is also my desire, that you should keep me constantly informed of the condition of the country, and of anything which occurs in it. You must therefore organize a postal service for the Gheziré (Sennaar), Kordofan, and Taka, from Gheziré to Abu-Khama. At each

interval of ten hours' march by camel, or about five
hours' by dromedary, you will establish stations for
two dromedaries, the riders of which will hand on
the despatches from one to the other. You will get
ready sheds in which they can be kept, and you will
provide means for feeding the messengers as well as
their dromedaries. You will establish three stations
between Abu-Khama and Korosko—the first at Abu-
Khama, the second at·Marat, and the third at Korosko,
so as to facilitate the arrival of your despatches. You
will also provide ten dromedaries for the mudir's
service.

" ' If in the event of any one of you being compelled
to assume the offensive, and of his enemies being so
numerous that he requires help from Cairo, send
me word at once, and I will send him the where-
withal to make their hearts faint within them, to
destroy and to disperse them ; and I will myself come
and punish those who have created disturbance and
done evil.

" ' Be well assured that the necessary preparations
will always be ready at Cairo, and that I will make
an example of those whom I find to be guilty. Be
convinced also that if I learn that the inhabitants
have been oppressed by you or by the sheiks, not one
of you will be spared punishment. Lay this well
to heart and act accordingly, for such is my order
and will.' "

Second Order of His Highness.

" ' In the order which I gave to you for the regula-
tion of the tax, and for the carrying out of other
instructions, it is stated that the tax is fixed upon this
basis since the solar year 1272 (Zilkedje 1273), that
the sum which the inhabitants may have paid since
the beginning of the year till now was to be deducted
from this year's tax, and that, out of my love for my
people, you were not to claim from the inhabitants the
arrears due up to the end of the year 1271.

" ' But as all this was not very clearly explained,
and as the inhabitants of these countries are unin-
structed, I fear that they may think that the arrears
are still due from them ; so I issue this order to set
their minds entirely at rest, that their joy and happiness may be full, and I explain to them more clearly
my wishes.

" ' The sums which have been collected since the
beginning of 1272 until now will be deducted from
the tax of the current year, after the accounts of the
serafs (surveyors of taxes) have been closely verified.

" ' With regard to those who are creditors up to the
end of 1271, for the excess which they have paid on
the tax which they owed, although in equity this sur-
plus should be made good by the arrears, yet in my
justice I do order that my subjects lose nothing of
what is due to them, and therefore you will compensate out of the tax] of the current year all those

who are found to be creditors for such sums duly proved.

" ' It is also necessary to be well acquainted with the limits of each village, and to compel the sheiks and notables to respect these limits and to appoint proper guardians, who will be responsible for any murder or theft committed within the boundaries of their village, and who will be bound to produce the murderer or thief, failing which they will be held personally responsible. This is done with a view to secure the safety of the road, and to prevent them shifting the responsibility from one to the other, which would render the process of trial a very long one, and make it very difficult to discover the truth.

" ' You will therefore take the necessary steps to fix the boundaries of each village ; you will make the sheiks understand what a serious responsibility rests upon them. . . . .

" ' Up till now the thieves and murderers sentenced to penal servitude for life have been sent to the galleys in the Soudan ; if, instead of that, they had been removed to galleys far away from their families and villages, the knowledge of this would probably have prevented them from committing the crime. I have consequently decided that those who are condemned to penal servitude for life shall be sent to the galleys in Egypt to undergo their punishment, and that those who are condemned to a like penalty in Egypt shall be sent to the Soudan.

" ' The accounts were formerly submitted to the Governor-General. But now that each province is independent you will send your accounts every three months to Cairo.

" ' You will communicate the contents of this order to all the sheiks and notables ; you will make them well acquainted with it, so that they may conform to it.

" ' Such is my will.' "

" Arakel Bey had begged me to ask the Viceroy to let him remain as Governor-General of the Soudan, 'so I took this opportunity of pointing out to him that it was no use to have good laws unless they were administered by suitable persons, and recommended Arakel Bey to him. The only objection he raised was that he feared the climate of the Soudan might be fatal to him, and he urged me to point this out to him, and to say that during the last few days low fever had killed half of the seventy Albanians who formed his escort.

" Despite this, Arakel Bey told me that he was anxious to have the honour of carrying out the noble ordinances of the Viceroy, and that it was the height of his ambition to be entrusted with this important mission. We were encamped near Khartoum, and upon my communicating Arakel Bey's decision to him, the Viceroy, who was always very prompt in his actions, at once sent for his ministers and generals, and addressed them as follows :—' You are aware that we are about to quit this terrible country, the climate

of which has already cost us so many valuable lives.
You are all men who have been enriched by my predecessors and by myself; you have palaces at Cairo;
you have families and every comfort; there is not
one of you who would have been foolish enough
to ask me to leave him here as governor of a country
which has been ruined. Well, the only one who has
aspired to this post is a Christian, Arakel Bey; he
really wants a straight waistcoat.'   Then one of his
ministers, Hassan Pasha, acting as buffoon of the Court,
seized Arakel and went through the pretence of tying
him up to the pole of the tent.   When this scene was
at an end the Viceroy made a sign and every one
withdrew, leaving us alone.

"'Well,' he said, speaking in excellent French,
'*le tour est joué*' (the trick is played). 'If I had been
compelled to appoint a personage in my train to act as
deputy for me in this important Government, with all
the external signs of my authority, my own tent, my
horses, my carriages, my palace, and all my absolute
powers, at a distance of six hundred leagues from my
capital, it would have been impossible for me to fulfil
my promise, on account of Arakel's religion, as there
is no precedent in the whole Ottoman Empire of a
Christian having occupied a like position.   Now you
can go and tell Arakel that his request is granted, and
that he can come and see me.'

"*January* 19.—I go to see M. Heuglein, whose
geographical information about the interior of Africa is

full of interest, and I meet there M. de Malzac, who had arrived the day before from the Upper Nile. He had been secretary to Count de Rayneval, French Ambassador in Rome, and he had abandoned diplomacy for the adventurous and perilous life of an elephant hunter in the Djours' country, between the 6th and 7th degree, ten days march inland, to the west of the White Nile. His cargo of ivory will bring him in about £1,600.

IV.

"Upon the 20th of January the Viceroy orders preparations to be made for a start, and we are to commence the journey in a week, traversing the vast desert of Bayuda, on the left bank of the Nile, as far as Dongola. This desert is much less inhospitable than that of Korosko, and we are to follow at the foot of the lofty mountain chain a series of valleys which are well cultivated, watered, and inhabited. It seems indeed as if this vast tract, described as a 'great desert,' upon the map, is not a desert at all.

"In the meanwhile we propose to make an excursion of two or three days up the Blue Nile, five or six leagues above Khartoum, to visit the ruins of Sheba, an ancient city of Ethiopia, perhaps the capital of the famous queen whom Solomon wished to have as his 301st wife.

"I advise the Viceroy to send for horsemen from a tribe in the province which, as I had been told, had armour and equipments for their horses similar to that

used by the French crusaders. Carriers were sent
out on dromedaries and soon returned, bringing with
them a dozen horsemen arrayed in coats of mail and
helmets, carrying long swords, the hilts of which
were in the shape of a cross, and riding horses richly
caparisoned from the head to the tail with very gaudy
cloth on a thick backing of cotton. They performed
some very clever feats of arms in our presence.

"Towards the end of January the Viceroy started
in advance of my caravan, as it was desirable not to
have too many people together in case the supply of
water at the wells should run short; but we arranged
to meet from time to time at certain halting-places
fixed before starting."

To Madame Delamalle.
(Continuation of the Diary.)
"CAIRO, *March* 6, 1857.

"I have at last arrived here safe and sound after
my long journey, having done the distance from Khartoum to the second cataract, which is about nine
hundred miles, in twenty-two days on an excellent
dromedary, which, however, was so tired during the
last week that he made a great many tumbles, and
tried my gymnastic abilities very highly. A steamer
was waiting for me at Ouade-el-Alpha * (the second
cataract), and the reason why I am a week behind the
Viceroy is that I was obliged to stop awhile in Dongola

* Note of the Translator.—Wadi-Halfa, as it is better known to
English readers since the Soudan campaign.

and attend the doctor whom he had told off for me. The doctor was very ill with low fever, and despite my want of experience in medicine, I succeeded in bleeding him and bringing him round.

"You are aware that, instead of returning by Korosko, upon the right bank of the Nile, we changed our itinerary so as to avoid the windings of the stream and five of its cataracts, and that we took the other route on the left bank of the river, through the so-called desert of Bayuda. I did not meet with a single accident or adventure in the course of this journey through a land occupied by supposed barbarian populations. Upon quitting the banks of the Nile and making for the country to the south-west of Khartoum, we traversed the tribe of the Hassanieh, the women of which, who are very handsome, are allowed complete liberty one day out of four.

"My caravan was always well supplied with provisions, while that of the Viceroy, which preceded mine, often ran short. The Prince asked me once how this was, and I answered him as follows : 'This is not at all to be wondered at. Your Government has so maltreated this country that, after you have passed through, I have to be very patient before I can overcome the mistrust of the inhabitants. Seated alone in front of an abandoned hut, and, letting my caravan get well out of sight, I have to wait an hour, or perhaps two, before the children will come near me. Children are always sent on in advance to reconnoitre.

If they hesitate to approach me, I throw them some small coins, some shells, or glass trinkets. They are sure then to go and tell their mothers what they have seen, and then the women come up, not as a rule the young ones. They surround me and ask me why I have made presents to the children, and I reply that I am a man of ease travelling for my pleasure, and for the good of the country in which I am sojourning. Then they all ask me at once if there is anything that you want. I tell them that, on the contrary, if they require provisions I have plenty at my encampment, which is an hour's march, and to which I invite them to come. It is when one has the appearance of requiring nothing that everybody is ready to furnish you with what you really do want. As soon as the old women had gone to fetch me the provisions, the young women and girls arrived, full of curiosity, very pretty some of them with their complexions like Florentine bronze, and they were soon followed by the young men. In short, a whole crowd of them came to our tents with sheep, goats, dates, and milk, and all that we could require. Curiously enough, they would never take any money, and yet these very same people would perhaps have killed me if I had come to them armed.'

"Another day the Viceroy said to me : 'You are very lucky, it seems. I had a fine service of china, but it is broken to bits.' I told him that if his china had been entrusted to men who were better looked after, it would not have happened. Soon after this he pretended that the camel which carried mine was tired out, and when the frisky one which he had put in its place kicked up and broke the handsome service, a gift of his own, he was delighted. Fortunately, I had in reserve what I call my silver service, made of tin and used by me while surveying for the canal, even when princes do me the honour of accepting my hospitality. To-morrow I am going to rejoin the Viceroy at one of his residences upon the Damietta branch of the Nile.

" It will be as well to give here the memoir which was read at the meeting of the Académie des Sciences in Paris on April 27th, 1857, by M. Elle de Beaumont, and which embodied my observations relating to the Soudan. These observations, which I put in the form of a letter, were as follows :—

" '*Monsieur Le Secrétaire Perpétuel,*

" ' Having received during my stay in Khartoum last January the questions and instructions of the Académie des Sciences, drawn up for the use of travellers seeking the sources of the White Nile, I communicated them to the Europeans, who were staying in or passing through Khartoum, and handed a copy to Arakel Bey, the Governor-General of the Sennaar provinces, who, by his education, fine feelings, and real worth, will not fail to exercise over these still barbarous countries a most salutary influence. I

requested this high functionary of the Viceroy to establish at his residence in the capital of the Soudan, and
in accordance with the questions sketched out by the
Académie des Sciences, a standing inquiry among all
travellers, tourists, savants, traders, and pilgrims,
whether native or European.

"'Circumstances favoured my commencing an inquiry
of this kind myself, and I had several opportunities
during my three weeks stay at Khartoum to question,
either together or independently, MM. de Malzac,
Thibaut, and Vayssières, French travellers; another
of our compatriots, Dr. Peney, who has been living
for the last ten years in the Sennaar; M. Heuglein, the
Austrian consul, and a very learned geographer and
naturalist; and Don Ignacio Knoblecher, the worthy
chief of the Apostolic Mission in Eastern Africa.

"'I am very pleased to lay before the Academy the
results of my investigations, and trust that they may
be deemed of interest.

"'Since the expedition of M. d'Arnaud, which did
not get beyond 4° 42' 42", no one has been further up
the river than Don Ignacio Knoblecher, Don Angelo
Vinco and Don Bartholomeo Mosgan. These hardy
missionaries navigated for a period of a fortnight
beyond the point reached by M. d'Arnaud, that is up
to the third degree. They formed at Gondokoro, in
the land of the Barrys, in 4° 35', latitude north, and
28° 47', longitude east (M.P.), an establishment which
is still flourishing despite the death of its first founder,

Don Angelo Vinco, and which is now almost as important as the mother house at Khartoum.

" ' The mission has lost within the last eight years twelve of its members out of thirty-six. It is at Khartoum that the climate is the most fatal to foreigners, owing to the prevalence of low fever. In 1839 Mehemet Ali lost in a week thirteen of the sixty persons who accompanied him, and the Viceroy the other day lost half of his escort of seventy Albanians who were encamped outside the town. They all died in the space of three days, during which the sun had been very hot.

" ' The outskirts of Khartoum need being drained, as the stagnant water which accumulates in the low ground after rain is the chief cause of mischief to Europeans. The city, founded by Mehemet Ali forty years ago, has now between 35,000 and 40,000 inhabitants. It is the centre of an important trade, and the very wise arrangements which the Viceroy has just made will certainly add to its salubrity and prosperity.

" ' M. Heuglein has ascertained it to be 1,060 feet (French) above the level of the sea, and its latitude you know. Khartoum, in Arabic, means elephant's trunk, and the name is derived from the comparison of the two branches of the Nile which meet here being like the two cartilages or snouts at the end of an elephant's trunk. The waters of the two rivers do not mix directly after their junction, those on the

eastern side being for some distance clear and blue,
while those to the west are muddy and of a whitish
hue.

"'In going up the White Nile from Khartoum to
the 10th degree the bed of the river is very broad
and slopes but very little, the result being that its
current is very slow, little more than half a mile an
hour, while with a north wind there is scarcely any.
The banks are not at all steep, and are formed by a
narrow sort of shore which divides the river from
the immense plains which are in many cases below
its level. The land is very well cultivated near the
river, but beyond it is covered with wild plants,
woods, and bush. At the 14th degree begins the
Archipelago of the Chulucks, up to within a day's
journey of the mouth of the Saubat, an affluent
running from the east between the 10th and 9th
degrees.

"'From the 10th to the 6th degree the White Nile
flows through marshes where travellers are much
plagued by insects. M. de Malzac, who last year
killed seventeen elephants with his own gun, has
formed an establishment in the land of the Djours,
between the 6th and 9th degrees, a hundred leagues
to the west of the Nile. From that point he has
put himself in communication with several other
tribes, all of which speak different languages. He
already employs five native interpreters to conduct
his exchanges of glass and other trinkets for ivory,

and as his relations are extending every year, he told me that he should soon require at least five fresh interpreters. To illustrate the necessity of this, he told me that an elephant is called *akou* by the Kilches, *keddé* by the Djours, and so on. Yet all these tribes have one word for the serpent, and that is *python*, the coincidence with the Greek being somewhat singular.

" 'A short time ago five hundred blacks came with M. de Malzac from his station to the banks of the Nile, carrying on their backs a cargo of elephant tusks which he was bringing down to Khartoum. This journey lasted a week, and the men passed over marshy land which beasts of burden could not have traversed. M. de Malzac had informed his men before he engaged them that as his stock of glass and trinkets was exhausted he could only pay them on his return. But this did not prevent them coming down to the river with their heavy load, and from returning home full of confidence in his promise.

" 'A fact like this shows that the inhabitants of these countries are not by nature hostile to strangers. Most of the tragedies which have recently occurred are due to the greediness and, in some cases, to the actual cruelty of certain traders.

" 'The *Niébor*, called in the Soudan the "Bahr-el-Gazal (Stream of Gazelles), is not, according to MM. Malzac and Veyssières, the principal part of the Nile, but only one of its affluents, and perhaps the most important. Among the Dinkas and the Chulucks the White Nile is called Kyr, and among the Barrys the *Churifiry.* Father Knoblecher states that when going up the river beyond Gondokoro he noticed upon the left bank at 4° 9' a granite mountain 500 feet high, which the natives call *Logouat.* While he was going up this mountain he felt a sharp shock of earthquake. The negroes who accompanied him, throwing themselves upon their faces to the ground, were very much terrified, and exclaimed that the spirits of the dead were coming back. Father Knoblecher having asked them what they meant by these spirits, they told him that there had formerly been a great battle in the neighbourhood, that the dead had been buried at the foot of the mountain, and that ever since their souls made occasional efforts to escape. The missionary took the opportunity, while combating their prejudices, to explain to them that the notions of the immortality of the soul, which they asserted were unknown to them, in reality came natural to them, and that it would never occur to them that the spirit of an ox or an ass could survive.

" 'A few lights to the south of Mount Logouat, on the right bank of the river, is a stream which is navigable for three days' journey, and which appears to have its source at the foot of a lofty mountain called Lologouchi. Further on, eight leagues from Logouat, commence the rapids, which are studded with islets and which extend for a hundred leagues, over which distance the river is not navigable. Father Knoblecher managed to pass through the first islets, but he was obliged to go on foot to a rock which is a hundred feet high, and from this elevated point he traced the Nile, as far as the eye could reach, flowing southward between two tall mountains called *Merek-Rego* and *Merek-Wigo*. It would appear from what is related by him and all other travellers, that beyond the rapids the river again becomes navigable as far as the 4th or 5th degree of south latitude, and that there it forms a bend towards the east, afterwards coming back towards the north, and having its source between the 1st and 2nd degree of latitude south, at the foot of a large chain of mountains called by the Somalis *Kœnia*, the tablelands of which nearest to the sources are called by the natives *Kali-Mandjaro*, or White Mountain. These, then, would be the silver-capped mountains, or the mountains capped with eternal snow, described by the Monbaz Protestant missionaries, as well as by the English navigator Short, who came from Zanzibar.

" ' Along the course of the White Nile, at the point where the rapids are met with, the two banks of the river are so close that the natives say they can shake hands across them. The Catholic missionaries have remarked that at several points a large tree is thrown across the river by way of a bridge.

"'The rising of the river begins in February or March. Sometimes the river will rise and fall again within the twenty-four hours, and this was what happened to Father Knoblecher when he was passing between the islets of the first rapids. He was afraid on one occasion that he should not be able to get back to Gondokoro, as his boat had stranded ; but the next day the water rose and floated it, this movement of ebb and flow occurring several times in succession. The Barrys, amid whom is situated the Catholic establishment of Gondokoro, belong to a numerous and powerful tribe, which is descended from a chief named Zangara, and from his sons, Karchiouk, Bepo, Pilza, Wany, Watavy, and Manabour. They were formerly in regular communication with a very distant tribe inhabiting the south-east, but the caravan which used to come to them every year has not been seen anything of for several seasons, owing to the hostile attitude of the intermediate tribes.

"'Ethnology.

"'The population is very dense all along the course of the White Nile wherever the land is productive. The arms used are lances, darts, large double-edged swords, ebony clubs, and tridents with three sharp blades, which the natives project with the hand. I send with this one of these tridents for the Academy's inspection, and two spades manufactured by the Djours out of the iron of the country.

" ' None of the tribes are able to write. They can count, and their system of numerals is similar to ours; and I append a tablet of the numerals, as supplied me by M. de Malzac, in use among the Kidgs, the Ajars, the Ocools, the Dinkas, &c.

" 'The dwellers along the White Nile live principally upon cow's milk, doura grain, sweet sorghhum, rice, beans, earth-nuts, and sweet potatoes. The married women are partially clothed in sheep-skins, but the men as a rule go quite naked. The Djours, however, enclose their generative organs in a panther-skin bag, while the women wear a belt of leaves round their loins.

" The habitations in the rainy regions are round huts with conical roofs; in the regions where no rain falls they are square, and have flat roofs. The Barrys invoke a divinity whom they call the great rain (*Dendit*). At a time of drought they sacrifice a white ox in order to obtain rain, and when there is too much rain they sacrifice a black ox to obtain sunshine. This sacrifice is, moreover, in very general usage among the tribes of the White Nile.

" ' When two enemies become reconciled, each of them puts to his lips a piece of iron, which is the token of peace, and which is at once buried in the ground at the spot where peace was made.

" ' The bodies of all those who die are cast into the Nile by the tribes who live on the banks; but the dead of the inland tribes are buried in front of their houses, in a sort of sitting position, which is only made possible by breaking the thigh-bones after death. A lance is thrust into the ground to indicate the tomb of a man, while on the tomb of a woman is placed the vessel which she has used for bruising the doura seed.

" ' Every evening the people meet to dance and sing. The singing is not so monotonous as that of the Arabs ; the tunes are lively and varied, and the singers have as a rule pleasant voices and keep time.

" ' Although the law is that of the strongest, the manners are for the most part very gentle. Theft and murder are rare, except in time of war, between family and family, or between tribe and tribe. Robbery is punished by the person who has been robbed, murder by the family of the victim. The leader of each tribe, the chief man of the family, is the one who is richest —that is to say, who has the most wives and stock. Polygamy is universal ; prostitution does not exist.

" ' The people consult soothsayers to obtain rain or heat ; but the calling is not always a lucrative one, and if the predictions do not come true the soothsayer is sometimes put to death by having his stomach opened. It will be easily believed that the sooth-sayer does not always await the return of his cus-tomers when his predictions have not been realised, and that he loses no time in disappearing when he is likely to be called to account in so shocking a fashion. The only public trade is that of blacksmith.

" ' ANTHROPOLOGY AND ZOOLOGY.

" ' I consider that from the Mediterranean to the fourth degree, and even farther, the populations along the Nile banks descend from races in which all the races foreign to Africa have been absorbed. The populations belong to two types quite distinct, but which are in some instances fused in the same locality, the Ethiopian and the negro types. The Ethiopian type dominates up to the tenth degree, but beyond that one encounters only the pure negro race, with its thick lips, flat nose, and woolly hair.

" ' It has often been asked if the Ethiopian populations have degenerated. I believe myself that they have remained stationary. They were probably during the splendour of the Egyptian and Ethiopian kings what they are now. It is the might of the kings and of the great which has perished with their palaces and their monuments. If you except these, with the royal tombs hewn in the rock or elevated on the pyramids, the private dwellings, the manners, the customs, the furniture, the arms, and the clothing were the same that they are to-day. The study of the monuments of ancient Egypt led Champollion to the conclusion that the valley of the Nile derived its first inhabitants from Abyssinia and the Sennaar, and that the ancient Egyptians belonged to a race of men very similar to the Barabras who inhabit Nubia at the present day. Diodorus of Sicily was also of that opinion, remarking

that even in his day the Ethiopians affirmed that
Egypt was one of their colonies.

" 'The tribes of the Upper Nile still plait their hair
as the ancient Egyptians did theirs. The sandals
found in the Egyptian monuments are the same as
those still used by the natives, and this holds good of
the wooden head-rests, the lances, the javelins, and
the shields.

" 'The children are comparatively light-skinned at
birth, the colour gradually deepening. The age of
puberty commences at about twelve or thirteen, and
the women do not bear child after they are forty.
The peculiarity of confinements in the Sennaar country
is that the women are placed in an upright position
against a wall, and that they are often suspended by
ropes passed under the armpits, and swung to and
fro or well shaken.

" 'None of the travellers or natives whom I have
consulted has ever heard of any men having a salient
coccyx.

" 'I have heard of some fellatah tribes of a swarthy
or reddish colour, supposed to be of Malay origin, and
living to the south and west of Darfour.

" 'I shall have the honour of presenting to the
Academy very shortly, on the part of M. Heuglein,
the complete notice which he has promised me on the
zoology of the White and Blue Niles. In the meanwhile, I append to this a manuscript map showing
the routes followed by M. Heuglein in his recent

voyages along the Nile in Abyssinia. M. Heuglein is a very keen observer ; he uses the most improved instruments, and he may be fully trusted as regards all the geographical points which he has fixed. He verified the absolute accuracy of the geographical observations of Bruce, especially with respect to the position of Lake Tana, which is traversed by the Blue Nile and just below its source.

" ' HISTORICAL REMARKS UPON THE EMPIRE OF MÉROÉ.

" ' No one has ever yet been able to say what was the extent of this empire, so rare are the remarks of ancient authors upon this subject. According to M. Heuglein, who has studied the question very closely upon the spot, the ancient Empire of Méroé was the Sheba of Scripture. It comprised Upper and Lower Ethiopia—that is to say Abyssinia, the Peninsula of Sennaar between the Blue and the White Nile, the Kordofan, the Peninsula of Méroé, between the Nile and the Athara (Astaboras), the provinces of Berber and of Dongola with Taka. . He derived this opinion from the inscriptions of Axoum and during his investigations of Ethiopian monuments. He discovered pyramids at six leagues from Rosérès (Sennaar, Blue Nile), at Debbah, and at the mouth of the two tributaries of the Blue Nile, the Yabous and the Taumat, to the south-east of Fazoglu.

" ' Besides the ruins of Méroé, discovered by Cail-

land in 1819, M. Heuglein has pointed out the existence in the peninsula of those of Ouad-Benaka, Wady-
Safrah, Wady-Okateb, of Sheba, the royal city on the
right bank of the Blue Nile, five leagues from Khartoum, and those of Khamlim ten leagues further inland
to the east.

" ' M. Heuglein has shown me a pen-and-ink map
which was recently sent him by Mr. Rehman, a Protestant missionary residing at Moubar, on the Zan-
guebar coast. This missionary appears to have
collected a good deal of information about an inland
sea called *Uniamesi*, of which there has been no little
talk recently, which is said to occupy an area of from
twelve to thirteen degrees north to south, and which
would in this case be larger than the Black Sea.

" ' The existence of this sea was certified to me
during my stay at Khartoum by a pilgrim from
Mecca, who inhabits Central Africa, and who gave
Mahmoud Pasha, one of the Viceroy's ministers, particulars corresponding to those upon Mr. Rehman's
map. This pilgrim added that he had seen larger
vessels on the *Uniamesi* than that in which he had
sailed down the Red Sea.

" ' I beg to place before the Academy a specimen of
india-rubber from the Djours country, which was
brought me by M. de Malzac, and this is, I think, the
first which has been discovered in any part of Africa.
I also send a fragment of colossal *convolvulus* which
sometimes reaches a length of thirty feet, a new

species of convolvulus named *djaugal,* which grows horizontally underground, and some *convolvulus gnocchi* growing upon stems, a kind of bean called *mangha* and fruit of the butter tree. These three kinds of convolvulus taste, when cooked, like our potato.

" ' MEDICAL PART.

" ' Dr. Peney, who has collected some very interesting information during his long residence in the Soudan with regard to the maladies prevalent in the country, has undertaken to prepare a medical treatise in reply to the questions raised by M. Jules Cloquet in his report of November 10th, 1856, and this treatise will be presented to the Academy. I may in the meanwhile communicate to the Academy a copy of the ordinances issued by the Viceroy for the reorganisation of the Soudan provinces, for these ordinances, so sensible and so liberal, while settling many important points, also bring to light a number of details relating to manners which are of a nature to interest the Académie des Sciences and which have a bearing upon several of the ethnological questions which are mentioned in its instructions.

" ' It may be said without exaggeration that from the issuing of these ordinances civilisation has been established and is feeling its feet in these remote countries, from which it seemed for ever excluded. I do not dwell upon the political consequences which these measures may have for the people to whom they apply. I only refer to the more or less scientific consequences. It is clear that the centre of Africa, hitherto almost inaccessible, will be much less so in future. The starting point will be Khartoum, placed beneath a Christian governor at the sixteenth degree, instead of Alexandria or Cairo, and it may be taken for granted that in a near future great explories will be made and great discoveries will be the infallible consequence. The researches, rendered more easy, will bear more fruit. Commerce will gain not less than science, and everything will be ready for a vast development of these fertile countries when the opening of the Suez Canal brings the coasting vessels of the Mediterranean into the Red Sea, and especially along the east coast of Africa. In these various ways the ordinances issued by Mohammed Said at Khartoum on the 26th of January open safer and more speedy roads to science, while they at the same time mark a decisive era in the amelioration of those lands.' "

" RESIDENCE OF THE VICEROY AT MIT-BIRÉ,
" (DAMIETTA BRANCH),

" *March* 7, 1857.

" His Highness was awaiting me at Mit-Biré, where we at once set to work giving orders for the continuance of the preparatory investigations and surveys. During our absence all the orders had been duly carried out, and as the master was absent no one dared say a word. Captain Pheligret, employed to take soundings in the Gulf of Pelusium, between the Damietta branch and the ancient Pelusian branch, did his work admirably. His vessel, despite the bad weather, held very well in the bay with only one anchor, and I intend to publish his observations.

" The course of the sweet-water canal has been carefully considered by Conrad and Linant Bey, and the plans are finished. The Viceroy is once more full of hope, and no one has attempted to shake this confidence.

" It appears that he has spoken to his family about my showing him real affection, for the princess his wife has thanked me in a letter written me at her dictation by Madame Stephan Bey, wife of the Minister of Foreign Affairs. Here is my answer :—

" ' *To Madame Stephan Bey, Cairo.*

" Mit-Biré, *March* 7, 1857.

" ' I told you when passing through Cairo how deeply grateful I felt for the gracious message which you were charged by the Vice-Queen to transmit to me; but I avail myself of the first moment which I can command to express to you my thanks in writing. Nothing could be more flattering than to receive this mark of high esteem from a princess known not only in Egypt but throughout Europe for her elevated character and intelligence, as well as for her acts of kindness and charity.

" ' What touched me most was to find that my feelings of devotion towards the prince, who has since his boyhood honoured me with his friendship, are appreciated by the person who would be best able to divine their nature, for gifted women have an almost supernatural instinct for picking out, almost without having seen them, the friends or the enemies of those to whom they are attached. Their views are rarely mistaken ; and there is no man, of those blessed with a faithful and disinterested companion, who has not occasionally had cause to regret not having followed the advice or given heed to the presentiments to which his vanity prevented him from paying attention.

" ' The Viceroy deigned to speak to me, during our voyage to the Soudan, of the high opinion which he had of the clear and straightforward judgment of his august spouse. This gives me a reason the more for rejoicing in the confidence which she is pleased to place in the sincerity of my attachment for a prince who may count upon ever receiving from me the free and respectful affection which his goodness of heart and, as I may venture to call it, his fraternal affection cannot fail to elicit.' "

Note to His Highness the Viceroy.

"Mit-Biré, *March* 9, 1857.

" As I count upon returning very shortly to Egypt, I would ask of your Highness to provide Linant Bey and Mougel Bey with the means for continuing the preparatory works upon the sweet-water

canal, in accordance with the plans agreed upon with
M. Conrad, President of the International Commission.
The number of workmen, which is now four hundred,
can then be gradually raised to a thousand, pending
the date for commencing the main works, which will
be fixed later on. It will also be advisable to get
together the material and the tools, of which a list
has already been drawn up ; and no time should be
lost in arranging for the making of bricks, the
excavation of stone, and the supply of wood."

To the same.

"PARIS, *March* 31, 1857.

"Upon my arrival I had the honour of an interview with the Emperor, and informed him that I was
not yet in a position to solicit the support of his representative at Constantinople. I was also able to give
him many details, which he listened to with much
interest, about your Highness's journey to the Soudan,
and the excellent results which would accrue from it.
The documents relating to the measures which you
decreed have been published here, and have been
made the subject of very favourable comment.

"I then proceeded to London, where I found that
the Suez Canal question had, in the course of the last
few months, made extraordinary progress. The leading merchants and bankers of the city received me
most cordially, and gave me letters of introduction
to the principal merchants, manufacturers, and ship-

owners in the fifteen largest towns of the kingdom. The Chambers of Commerce, the merchants, the manufacturers, and the shipowners of these towns have been informed that I am going to commence a series of visits to them all about the middle of April, and nothing will be left undone to render this tour decisive of the question so far as England is concerned. My object is to collect signatures and declarations to the effect that the piercing of the Isthmus of Suez will be beneficial to English interests, as well as to those of other nations, and that no government has any right to put obstacles in the way of the work.

"In this way your Highness's glorious enterprise will be based upon public opinion in England, as it already is upon that of the European continent and America. While using all my efforts to attain that end, I do not forget my promise—I may add, my duty—to avoid anything which might be calculated to disturb your Highness's friendly relations with all the Powers.

"After what I have myself seen in Paris and London, and from what M. de Negrelli writes me from Austria and Signor Palescopa from Italy, everyone praises your Highness for having commenced the sweet-water canal; and I can confidently assure you that you can continue the work without the least cause for uneasiness, if the weather, the requirements of agriculture, and the government resources admit of your doing so.

" In any event, your Highness is certain to decide for the best; and when my English tour is ended. and I am prepared to go to Constantinople, I will first come to Egypt to take your orders."

Meetings.

The months of May and June, 1857, were devoted to going to the principal towns in England, Scotland, and Ireland. The resolutions passed at these meetings were unanimously in favour of the execution of the canal, that which was carried at the London meeting (June 24th, 1857) being similar in terms to the rest:—

" At the public meeting of merchants, bankers, shipowners, &c., held at the London Tavern, Wednesday, June 24th, 1857, Sir James Duke, Bart., in the chair, it was proposed by Mr. Arbuthnot and seconded by Captain Harris, of the P. and O. Steam Company, ' That the canal through the isthmus of Suez having been declared practicable by competent engineers, and all nations having been invited to take part in the enterprise, which will not be placed under the exclusive protection of any government in particular, this meeting, being quite satisfied with the explanations given by M. de Lesseps, is persuaded that the success of the canal will be eminently advantageous to the commercial interests of Great Britain.' Carried unanimously.

" James Duke, Chairman."

The account of all the meetings, beginning with that at Liverpool on April 29th to that at London on June 24th, was published in English, and it was dedicated to the members of the Houses of Parliament in the following terms:—

"I dedicate to you individually, and I submit to your illustrious assemblies, the following pages, which embody the resolutions and deliberations of the principal towns in the United Kingdom, the commercial and municipal corporations of which have formally expressed their opinion upon the interests of the trade, the navy, and the colonies of Great Britain, as they would be affected by the opening of the canal through the isthmus of Suez.

"Reassured as I now am as to the competent opinion of the traders, the manufacturers, and the shipowners of Great Britain, and being about to pursue the execution of the work upon behalf of which I do not ask for the protection or the exclusive help of any government, I appeal in all confidence, in order to put an end to the opposition of the British Ambassador at Constantinople, to the political bodies of a free country which, in other circumstances, have already had the glory of placing above every consideration of private interests or national rivalry the great principles of civilisation and free trade. This pamphlet, addressed to politicians, would be regarded by them as incomplete unless I passed in review the elements of the political questions which have been raised in connection with the enterprise. It has been said that the opening of the African isthmus would threaten the power of England in India, and in this connection an effort has been made to revive the ancient distrust of England for France.

"The Suez Canal has also been represented as calculated to loosen the bonds between Turkey and Egypt, and to bring about the independence of the Egyptian Viceroy. Instead of avowing a hostility which it is no longer possible to conceal, this hostility was masked beneath such reasons as the so-called interests of Turkey, or was attributed to members of the Divan, who have repudiated it altogether, either in letters which have been shown to me or in their conversation with the representatives of the various governments which have not scrupled to express their unrestrained sympathy with the undertaking.

" Of these three questions of the relations between France and England relative to the Suez Canal, of the respective situations of Egypt and Turkey, and of the interests of Turkey in the piercing of the isthmus of Suez, the first was discussed in a letter which I wrote to Lord Stratford de Redcliffe at the outset of the enterprise, and the two others in the subjoined notes which I submit to the impartial judgment of my readers :—

" ' The enlightened Turks, far from being alarmed at them, see, upon the contrary, in the consequences of the opening of the Suez Canal a guarantee of security for the future. What they dread above all else is the risk of being exposed to any dangerous eventualities upon the part of one or other of the European Powers. They will always wish that Egypt should be exceptionally governed by Mussulman princes of Turkish origin, who are connected by so many common political and religious ties to the metropolis of Islamism.'

" With regard to the Viceroy of Egypt, in his communications with Turkish statesmen, speaking of the attempts made to raise a prejudice against him, he said : ' In the present state of things a ruler of Egypt who had any secret idea of aggrandizing his position would not allow the Suez Canal to be made. The whole of the coast, from Damietta to the first ports of Syria, is at present beyond the reach of any foreign surveillance, as it is outside European navigation. Nothing stands in the way of the Viceroy arming a fleet or collecting troops without exciting notice, and of throwing them into Syria before any one could interfere. When the canal is made the whole situation will be altered. Moreover, the important possessions of Turkey in Arabia can easily be reduced by starvation, as Egypt has the supplying of them with corn. There always exists in these provinces slight elements of rebellion, which it would be easy for Egypt to keep alive and increase, and which she alone, with the existing means of communication, could alone put down. Experience has shown that the distance and the difficulty of transport prevents Turkey from sending to Arabia enough troops to ensure her the preponderance of power. Then we are told that the canal would create a barrier between Turkey and Egypt. Anyone who knows the country must be well aware that, in a physical sense, a vast desert without water is a far greater barrier between them than would be the maritime and the sweet-water canals, around which large numbers of Syrian and Egyptian cultivators would gather.'

"This language is not less remarkable for its outspoken honesty than for its striking truthfulness."

DEBATE IN THE HOUSE OF COMMONS, JULY 7, 1857.

The Isthmus of Suez Canal.

Mr. H. Berkeley asked the First Lord of the Treasury whether her Majesty's Government would use its influence with his Highness the Sultan in support of an application which had been made by the Viceroy of Egypt for the sanction of the Sublime Porte to the construction of a ship canal across the Isthmus of Suez, for which a concession had been granted by the Viceroy of Egypt to M. Ferdinand de Lesseps, and which had received the approbation of the principal cities, ports, and commercial towns of the United Kingdom; and if any objection were entertained by her Majesty's Government to the undertaking, to state the grounds of such objection.

Lord Palmerston:—Her Majesty's Government cer-

tainly cannot undertake to use their influence with the
Sultan to induce him to give permission for the construction of this canal, because for the last fifteen
years her Majesty's Government have used all the
influence they possess at Constantinople and in Egypt
to prevent that scheme from being carried into execution. (Hear.) It is an undertaking which, I believe,
as regards its commercial character, may be deemed to
rank among the many bubble schemes that from time
to time have been palmed off upon gullible capitalists.
(Hear and a laugh.) I believe that it is physically
impracticable, except at an expense which would be
far too great to warrant the expectation of any returns.
I believe, therefore, that those who embarked their
money in any such undertaking (if my hon. friend has
any constituents who are likely to do so) would find
themselves very grievously deceived by the result.
However, this is not the ground upon which the
Government have opposed the scheme.   Private individuals are left to take care of their own interests,
and if they embark in impracticable undertakings they
must pay the penalty of so doing.   But the scheme is
one hostile to the interests of this country—opposed to
the standing policy of England in regard to the connection of Egypt with Turkey—a policy which has been
supported by the war - and the Treaty of Paris.   The
obvious political tendency of the undertaking is to
render more easy the separation of Egypt from Turkey.
It is founded also on remote speculations with regard

to easier access to our Indian possessions, which I need not more distinctly shadow forth because they will be obvious to anybody who pays attention to the subject. I can only express my surprise that M. Ferdinand de Lesseps should have reckoned so much on the credulity of English capitalists as to think that by his progress through the different counties he should succeed in obtaining English money for the promotion of a scheme which is in every way so adverse to British interests. (Hear, hear.) That scheme was launched, I believe, about fifteen years ago as a rival to the railway from Alexandria by Cairo to Suez, which, being infinitely more practicable and likely to be more useful, obtained the pre-eminence ; but probably the object which M. de Lesseps and some of the promoters have in view will be accomplished, even if the whole of the undertaking should not be carried into execution. (Hear and a laugh.) If my hon. friend, the member for Bristol, will take my advice, he will have nothing to do with the scheme in question. (Hear, hear.)

To the Members of the Chambers of Commerce and of the Commercial Associations of Great Britain.

"Paris, *July* 11, 1857.

" I cannot pass over in silence the assertions which the First Lord of the Treasury has thought fit to make with reference to the Suez Canal scheme at a recent sitting of the House of Commons. Replying to Mr. Berkeley, he expressed himself hostile to the making of the canal upon commercial, technical, and political grounds, making use of personalities for which I prefer not to seek an appropriate designation. With regard to the first point, that relating to the commercial advantages of the canal, I find an answer in the unanimity with which the eighteen principal commercial and industrial towns of the kingdom pronounced in its favour. You have been unanimous in declaring that this canal, abridging by one-half the distance to India, would be advantageous to British commerce.

" With regard to the second point, I answer Lord Palmerston by the mouth of the International Commission, composed of eminent engineers and mariners of all nations, England included, who, after two years of minute study and careful exploring of the ground, decided in the name of science that the making of the canal would be not only possible but easy. I answer Lord Palmerston with the sanction given to the opinions of the engineers and their plans by the Académie des Sciences in Paris.

" You will decide, gentlemen, between the authority which this verdict, emanating from the leaders of European science, carries with it and the unknown authority to which Lord Palmerston vaguely alludes. Without dwelling at length upon the contradiction involved in treating the project as chimerical, and at the same time denouncing it as dangerous, I come to the third point. The political arguments of Lord Palmerston seem founded upon the imaginary dangers which the Suez Canal would create for India, as well as for the integrity of the Ottoman Empire. The English press has already declared, of its own accord, that the masters of India have nothing to fear from the Mediterranean Powers as long as they are in possession of Gibraltar, Malta, Aden, and have just taken Perim. Turkey is at least as much interested as Lord Palmerston in seeing that Egypt is kept within the limits assigned to her by treaty. Now, the Divan is so far from regarding the canal as a cause of separation, that the English Ambassador is obliged to bring his full weight to bear in order to defer the ratification of the project. It is clear to the Porte, as it must be to all reflecting minds, that the opening of the isthmus, guaranteeing, as it will, Egypt against all foreign ambition, will add a fresh force to the integrity of the Empire, and be fraught for Turkey with religious and economic consequences of the highest importance.

" If a systematic yet unavailing opposition is persisted in, the enterprise may be beset with difficulties which will aggrandize rather than weaken it, but its execution will be resolutely gone on with, and the universal support accorded it will render its success infallible. In the meanwhile, it will be for the commercial classes of England to decide whether, in opposition to the views they have manifested, the obstacles are to be raised by their own Government. It will be for them to say whether they will allow a policy so

contrary to the principle of free communications and
free trade, which their nation has proclaimed in the
face of the world, to be carried out in their name, and
whether further efforts shall be made to prevent the
joining of two seas which lead direct to India and to
China, while in other ways they are doing all they
can to bring these vast countries into contact with
civilised peoples.

"I now come to the personalities, and I will endeavour, in replying to them, to observe the rules of moderation, considerateness, and dignity, which have
scarcely been adhered to by making an attack upon
me in an assembly where I could not be heard in
defence. Lord Palmerston thought fit to state, in
terms that I will not stoop to repeat, that I had come
over to England with designs upon the pockets of his
countrymen, and in order to take advantage of the
credulity of any capitalists who might be weak enough
to believe in a chimerical enterprise. You know,
gentlemen, whether I have said or done anything to
justify imputations of this kind. Have I made a
single appeal for subscriptions? You will remember
that, upon the contrary, I have several times told you
that I had come to ask you, not to subscribe for shares,
but for an expression of your opinion. If, in the
allotment of a capital of eight millions, England, like
France, is ultimately to have a fifth share, I made
this proposal out of deference to a powerful commercial nation directly interested in the opening of

the new route. But the enterprise of which I am the promoter stands so little in need of English capital that if the share allotted to England was not accepted in its entirety by her, it would be at once snapped up by demands coming from all parts of the globe.

"Such, gentlemen, is the simple and, as I believe, irrefutable answer which I have to make to Lord Palmerston, and which I address to the heart and conscience of all honest men. You will do me the justice of allowing that, in my reply, I have had proper regard to what is due to the age and political standing of the First Lord of the Treasury. I should, moreover, deem it inconsistent with my own dignity, and with the respect which I entertain for you, if I allowed myself to speak of him in such language as he has applied to me. I owe you these explanations because of the kind esteem you have shown me, and for which I feel profoundly grateful."

Note for the Emperor and Count Walewski.

"PARIS, *July* 15, 1857.

"I have the honour to enclose a letter which I have written to the British Chamber of Commerce, in reply to Lord Palmerston with reference to the Suez Canal.

"It had been agreed, as a matter of principle, that M. Thouvenel should be free to take action in favour of the canal in case Lord Stratford de Redcliffe should make any hostile move, but that, pending an agreement between the two Governments, their respective agents should maintain a neutral attitude with regard to an enterprise due to private initiative.

" Lord Palmerston now publicly declares that ' H. B. M.'s Government has, up to the present time, used all its influence to prevent the project of the Suez Canal being carried out.' In view of such an avowal, based upon inveterate mistrust of France—a mistrust which it is no longer thought worth concealing—need we really await Lord Palmerston's leave to make a formal demand upon the Sultan for the ratification of the Viceroy's act of concession, especially when we know that the Sultan is disposed to grant this demand? When we remember that the British Government, without troubling itself as to what an allied government might think of it, has obtained from Constantinople several important concessions, among others that of the Euphrates Railway, officially supported as being the English military road to Asia, and that it has recently seized Perim, a dependency of Turkey, without even so much as notifying the fact; and when we further remember that the opinion of the commerce of Great Britain is unanimous in favour of the canal, who could venture to complain if the representative of France was authorised to protect, in agreement with the representatives of the principal Powers who are in favour of the scheme, the interests of the holder of the concession, who is a Frenchman, and who has, moreover, but one interest to serve, that of opening a commercial route profitable to the whole world.

" I, of course, understand that the Imperial Government must choose its own time. I will await that time, going on in the meanwhile with the preparations for the project; and if the matter is allowed to drag on very long, all that will remain to be done will be to formally recognise an accomplished fact."

To His Highness the Viceroy.

"*July* 19, 1857.

" I beg to forward to your Highness the note which I have just handed to the French Government, and with it I enclose extracts from English newspapers referring to the debate in the House of Commons on the 7th inst. I am not called upon to say what I think of Lord Palmerston's language, which is severely condemned by several important organs of public opinion, among others *The Advertiser* (Bristol) and *The Daily News* (London).

" *The Advertiser* says :—

" 'The Isthmus of Suez Canal.

" ' Two great works have for some time been proposed to be undertaken. They would both, if accomplished, take the shape of grand ship canals, the one piercing the narrow strip of land that connects North and South America, the other slitting up the Isthmus of Suez, and thereby joining the waters of the Mediterranean with those of the Red Sea. The construction of the former is now more problematical than it was some years ago, the surface of the land having been found to be difficult, with many alternations of hill and plain. Circumstances may hereafter, in the pressure of commercial necessity, compel the work to be done, but at present interested speculators are content with patched routes, partly by rail and partly by water, from the Atlantic to the Pacific. The country which forms the Isthmus of Suez is understood to be much more favourable for the construction of a canal, and that operation many are hopeful will be carried to maturity.

"'If it be so, it will not be the first time that the isthmus has been channelled. A canal connecting the Red Sea with an arm of the Nile was commenced about 2,500 years ago, and was (according to Herodotus) completed by Darius. It is now as dry as the desert, although numerous traces of its ancient direction still appear in different places. The increased traffic with China in recent years, and the gold discoveries, and consequent expansion of commerce in Australia, have naturally caused the attention of inquiring minds to be directed upon any available means of shortening the distance between Europe and those distant lands; and, inasmuch as the projected canal across the American isthmus of Darien gradually fell into a state of quietude, it occurred to the mind of M. Lesseps, a French engineer, that the sandy plains

of the Egyptian isthmus might be so operated on as
to effect nearly the same object. Cut a ship canal
between the Mediterranean and Suez at the head of
the westernmost of the two arms or gulfs in which
the Red Sea terminates, and by a short water
route of 92 miles across the isthmus about 5,000
miles would be saved in the voyage between this
country and India, China, and Australia. Now,
could such a saving be effectually accomplished, the
advantages which it would confer on commerce would
be enormous; and shipowners and commercial men
generally should lend the project every aid of which
it is found to be deserving. It is probable that few
engineering difficulties would be experienced in cutting a canal through the isthmus, for the material to
be excavated consists generally of sandstone lying in
horizontal strata, or of sand, the consequence of dis-
integration of the sandstone. The main difficulty
would probably be found in the Red Sea, with regard
to its capability of allowing the passage of " the
largest ships " throughout its entire length of about
1,400 miles. We observe that at the meeting on the
subject held last week in Bristol, Mr. D. A. Lange
said " experiments had been made which showed that
the bed of the sea was singularly adapted for dredging,"
which countenances the apprehension that the waters
of " this sea " are in parts comparatively shallow,
however deep generally; and it will be only common prudence to ascertain all about the necessity of

" dredging " a sea before investing eight or ten millions sterling in the formation of a ship canal capable of accommodating vessels which might by possibility be stopped at Suez or somewhere in the long navigation that ensues before the Straits of Bab-el-Mendeb are left behind. To ascertain the actual state of the variable Red Sea should be a chief object of preliminary survey, for its navigation is as yet comparatively obscure, although the port of Suez is the point of communication between Europe and India in connection with the Overland Mail.

" 'The resolution moved by Mr. R. P. King, after stating that the projected ship canal would be of the greatest importance to the commerce of the whole world, added, "And would afford facilities which no railway could present." This is a cut at a rival scheme for shortening the route to India, and for generally facilitating the intercourse of Europe with Asia, which has been devised, we believe, by Colonel Chesney, who proposes to carry a railway from the Mediterranean into the valley of the Euphrates, to follow the course of that river south-eastward, and thence proceed to Hindostan by way of Persia and Belochistan. It really does appear that such an undertaking would be more formidable than cutting a canal 92 miles long through sand and sandstone. Much, however, as already said, depends upon the character of the navigation of the Red Sea—its winds, its coral reefs, &c. ; and if it be correct that M. Lesseps's project has received high engineering testimonials in its favour, it must not be forgotten that Colonel Chesney has carefully surveyed the entire route from the Mediterranean to the Euphrates, and the course of that river to the Persian Gulf, and is himself a practical engineer of the highest possible authority.

" ' Supposing that no insuperable material difficulties are found in the way of M. Lesseps, and that money is obtained to form the canal, a trade revolution would be effected calculated to surprise the world. In that case Europe need not care about the ultimate proceedings in the Isthmus of Darien, and the navigation of the stormy Cape would be almost forgotten. The resources of Arabia and Eastern Africa would be developed, as far as they are capable of development, and the voyage to India, Australia, China, &c., be shortened by about a third.

" ' We think, consequently, on the whole, that the merchants and shipowners of Bristol have done well to accord to M. Lesseps their frank and cheering countenance, as a preliminary, mayhap, to their pecuniary support. No national jealousy should exist in such a case. And if we have seen some ground for suggesting caution, we should have done the same had Colonel Chesney patronised the canal and the Arabian Gulf, and the French engineer had projected a railway through Asia Minor, and so on to the regions of the far East.' "

To Mr. Robert Stephenson, M.P., Engineer.

"LONDON, *July* 27, 1857.

" I enclose you a copy of the speech, as reported in *The Times*, delivered by you in the House of Commons on the 17th inst., and I shall be obliged if you will inform me whether this report is a correct one. The engineers of the International Commission, who have all their lives long devoted their studies to the construction of ports and canalisation, can best answer the technical part of your speech; but there is one point to which I venture to call your attention, because it concerns me personally. You said, according to *The Times*, ' I agree with the First Lord of the Treasury.' Now, Lord Palmerston, who holds a position which prevents me from addressing myself to him personally, had just spoken as follows :—' I do not think, therefore, that I am far wrong in saying that the project is one of those chimeras so often formed to induce English capitalists to part with their money, the end being that these schemes leave them poorer, though they may make others much richer.' I ask you, sir, for a written explanation of what you mean, either furnished by yourself or by two of your friends, whom you will please put in communication with me. I do not doubt that you will at once give me these explanations. I have come over from France on purpose to ask you for them. I have the honour, sir, to place myself at your disposal."

Mr. Charles Manby to M. Ferdinand de Lesseps.

"LONDON, *July* 28, 1857.

"Mr. Stephenson returned this morning, and I at once gave him your letter, which I had translated word for word. He repeated what, as I had already told you, he had said—viz., that his remarks about the canal were based upon the ideas he had formed in the course of his two journeys to the desert, and that he had only expressed his opinion in the House when appealed to by Lord Palmerston and several members who had your pamphlet in their hands. He has expressed his extreme regret that you should have supposed that he meant to make any attack upon your personal character, or that he endorsed any expressions of Lord Palmerston which might be taken to have this meaning. Upon the contrary, he has always held you in high esteem, and has invariably spoken of you in that sense.

"Moreover, he has gladly written you the enclosed letter which, I hope, will convince you that he merely expressed a technical opinion upon a matter being publicly discussed. Mr. McLean agrees with me that Mr. Stephenson had not the slightest intention of saying anything personally offensive to you."

Reply of Mr. R. Stephenson to M. F. de Lesseps.

"LONDON, *July* 28, 1857.

"Dear Sir,—Nothing could be further from my intention, in speaking of the Suez Canal the other

night in the House of Commons, than to make a single
remark that could be construed as having any personal allusion to yourself, and I am confident no one
who heard me could regard what I said as having any
such bearing. When I said that I concurred with
Lord Palmerston's opinion, I referred to his statement, that money might overcome almost any physical
difficulties, however great, and that the undertaking,
if ever finished, would not be commercially advantageous.

"The first study which I made of the subject, in
1847, led me to this opinion, and nothing which has
come to my knowledge since that period has tended
to alter my view.

" Yours faithfully,

" ROB. STEPHENSON."

*To Mr. Charles Manby, Secretary of the Society of Civil
Engineers, London.*

" LONDON, *July* 29, 1857.

" I have received your letter of yesterday, together
with that of Mr. Stephenson. While satisfied with
his explanations, so far as regards myself, I am still
very much astonished that an engineer should have
allowed himself to express himself in the House of
Commons so dogmatically with regard to an enterprise which he has not been in a position to examine
either upon the spot or in his study, especially when
he fails to give at the same time the grounds upon

which his opinion is based. The eminent engineers who form the International Commission will answer him in a very short time. He will then have to speak very explicitly upon the technical question, and I shall be very well satisfied if the ancient or recent studies of Mr. Stephenson shed any new light upon an enterprise which has for the last three years been under the attentive examination of all the *savants* in Europe."

To M. Barthélemy St. Hilaire, Paris.

"LONDON, *July* 30, 1857.

"I shall not leave London till I find that there is nothing more for me to do.

"I am thankful that I was not there when the questions were put by Messrs. Berkeley and Darby Griffith, as I could not have prevented them, and it would have been risky to have asked our supporters in Parliament to get up a debate when Lord Palmerston has so large a majority. This majority would, in order to keep him in office, have voted against us, which, as matters stand, it has not done, thus leaving Lord Palmerston alone responsible, in the eyes of Europe, for the use of language all the more violent and absurd because there was no one to answer him, and for a policy which is generally condemned, even in England.

"We had thought that it would be very advisable to get public opinion in France to express itself in some legal form with regard to the Suez Canal. The Councils-General are summoned to meet next month ; Lord Palmerston's attacks have stirred public feeling ; the French press of all shades of opinion, with true patriotic feeling, has strongly condemned them. We ought to take advantage of this state of things. I send you the draft of a circular, which might also be sent to the Chambers of Commerce, whom we will ask to pass resolutions in favour of our enterprise."

To M. Thouvenel, Constantinople.

"LONDON, *August* 2, 1857.

" After Lord Palmerston's declarations I am more certain of success than ever. When the time comes the financial co-operation of France may be counted upon without a doubt.

" No one here has ventured to stand by the First Lord of the Treasury ; he has been condemned by the leading men in the country, even by those who, in the critical position now occupied by England, think it their duty to keep him in office. I had been told of this by letter while in Paris, but I thought it best to come over here and satisfy myself that such really was the case. I may add that my own observations, to say nothing of the exceptional warmth of my greeting, fully confirm this view.

" I agree with you that the Constantinople press should be very prudent, and I have already urged my friends to treat the position of the Porte, powerless though it is, with the utmost tenderness and deference. But in due course the Divan will certainly, in presence of the universal wishes and support of other Powers, be bound to assert its independence and dignity before the world. I may add that the accomplishment of these duties will be a source of strength rather than of embarrassment. This is the opinion of Prince Metternich, one of the oldest and most trusty friends of Turkey.

"They must be beginning to see at Stamboul, especially since the seizure of Perim, that if a certain great Power wishes to close the Red Sea, as she succeeded in doing more than a century ago, by a decree of the Porte, it is with a view to her sole profit, and not in the interest of the Ottoman Empire, for whom rapid communication with the holy places of Arabia is almost a matter of life and death. It is not very long since *The Times* declared that Great Britain was 'the first Mussulman Power.' It was hitherto supposed that Turkey was. I know who wrote that article, and you may be sure that it was only a feeler. According to this system the seizure of Perim would be only the first step in a more complete invasion."

To His Highness the Viceroy.

"PARIS, *August* 12, 1857.

"The manifestations of the commercial bodies and of the citizens of all countries day by day condemn more strongly Lord Palmerston's declarations, but I cannot affect to ignore that these declarations, which will serve as a guide to the diplomatic agents of England, will cause your Highness a good deal of annoyance, which I should wish to spare you. You can put upon me all responsibility for the preliminary works on the canal, and with this view I have informed MM. Renaud and Lieussou, who have been appointed to survey for the making of the sweet-water canal, that I was about to propose to your Highness not to execute the work at your own cost, but to leave it in the hands of the Universal Company, which will doubtless be organized very shortly.

"If we look back to what occurred in regard to Egypt during the years 1839-40 we find that there is a good deal of analogy between then and now. Thus among the grievances alleged by the Porte, at the instigation of Lord Ponsonby, the English Ambassador, to justify the armed intervention against Mehemet Ali, was one to the effect that he had attempted to interfere with Great Britain's communications with India, by way of Egypt and Syria. The only foundation for this charge was in the following opinion, confidentially expressed by Mehemet Ali in a despatch to the Grand Vizier :—

"'That the opening of the passage from Europe to the Indies, by way of Egypt and Syria, ought to be made for the benefit and with the concurrence of all nations, and ought not to constitute a monopoly for the profit of England alone, a monopoly which would be very dangerous for the rights of the Sultan.'

"This question was referred to in the French Chamber, in the course of a debate upon the negotiations which followed the battle of Nezib, and M. de Lamartine spoke as follows :—

" 'Nature is stronger than these wretched national antipathies. Europe and India will communicate, despite all you may do, by way of Suez. You will but have delayed this great and beneficent act of Providence ; the two worlds will join hands, and gather new life as they do so, by way of Egypt.'

"We have now the Indian mutiny, which will supply the English press with a new and powerful argument against Lord Palmerston, and against the reluctance to make use of the route through Egypt. An Englishman writes as follows to *The Daily News* :—

" 'The *last* news of the mutiny in India reached England on June 17th. Since then a body of 2,000 men might have been despatched from England every fortnight, and have reached India by way of Egypt in six weeks. Why does not the Government send troops to India through Egypt? The Government has refused to answer. It is because of its reluctance to furnish the promoters of the Suez Canal with an argument the more.'

"In the meanwhile the mutiny is running its course, and costing the lives of many brave men, who were looking for more prompt relief than that sent by way

of the Cape. More than this, Nana Sahib, in a proclamation addressed to the Mahometans of India, tells
them that the Sultan, in a firman addressed to the
Viceroy, has ordered him to close Egypt, 'which is
the route to India,' to the British troops, that in
consequence there was no need to be afraid of their
approaching arrival, and that on receipt of this news
Lord Canning, the Governor-General, 'was overwhelmed with despair, and was beating his head.'

"This Indian insurgent little knew when he invented this piece of news that it was the reverse of
the truth, and that the able and enlightened ruler of
Egypt was preparing for the opening of the Suez
Canal, which the Prime Minister of England and her
ambassador at Constantinople were opposing.

"The English journal which publishes Nana Sahib's
proclamation adds, *Fas est ab hoste doceri.*"

To the same.

"LA CHÉNAIE, *September* 10, 1857.

"I forward to your Highness copies of the resolu-
tions addressed to the French Government by the
Councils-General and the Chambers of Commerce,
together with several letters of foreign Chambers of
Commerce, among which that of the Barcelona
Chamber deserves special mention.

"The English Government has at length made up its
mind to send troops to India through Egypt. Your
Highness is too high-minded not to favour in every

possible way the despatch of these troops intended to ensure the triumph of civilisation over barbarism.

"Lord Palmerston's conduct is still very severely condemned, and one journal says : ' Let us hope that he will see by this what a blunder he has made, and how dangerous it will be for him to persist in it.'

" But this is not all, for, in addition to the Councils-General, the Chambers of Commerce of the thirty-seven largest French towns have sent resolutions to the Government expressing their concurrence in the project for making the canal, while the Paris Chamber of Commerce has placed itself at the head of these manifestations which are only just beginning. With less obligation to be guarded in their attitude than the Councils-General, the Chambers of Commerce also protest against the attitude of Lord Palmerston, and urge the Government to intercede and ensure the execution of a project which will be one of the glories of the century."

To Mr. Darby Griffith, M.P., London.

" Paris, *September* 15, 1857.

" I have read with much interest the speech which you made in the House of Commons, and of which you have been kind enough to send me a copy.

" You expressed with force and eloquence the most noble and just ideas as to the true policy of England in this important question. I feel, like you, very certain that Lord Palmerston is making a most unfor-

tunate blunder in thus opposing a work which will
be more useful to British commerce than to all the
rest of the world. This course is all the more ill-
judged because it has no chance of succeeding, and
if, in the eyes of some politicians, the end justifies the
means, Lord Palmerston's conduct, in his deplorable
campaign against the Suez Canal, has not even the
chance of succeeding.

"Permit me to make some minor criticisms with
regard to certain details of your remarkable speech.
No doubt what you say about the workmen in Egypt
holds very true of the time when you were travelling
through the country. But since the accession of the
new Viceroy there has been a great change. The
cleaning out and the enlargement of the Mahmoudie
Canal in April, 1856, prove that at the present time
public works are carried out with due humanity, and
that the task set the workmen is neither beyond their
strength nor fatal to their health. Out of 115,000
men assembled for a full month, not more than five
or six per thousand fell ill. I doubt whether we
could show a better average than this in Europe. In
making the Suez Canal, it will be very easy to bring
the Nile water as far as Lake Timsah, in the centre of
the isthmus, which it reaches even now when the
river rises. This region, now barren and uncultivated,
formerly had a considerable population, and we discovered there the ruins of many cities. It was the
land of Goshen spoken of in the Bible.

" As to the practical difficulties, whether at Suez or Pelusium, they are not nearly so great as might have been imagined previous to the survey made by the eminent engineers who spent some time in the isthmus, and the very conclusive observations made in the Bay of Pelusium.

" To conclude, I may add that you seem to me to be too well versed in economic questions not to be led, after careful examination, to the conclusion that the enterprise will be financially remunerative, if you cast your eye over the official statistics which show how enormously European trade is increasing in Asiatic waters, the English figures for 1856 showing an increase of 181,000 tons over the previous year."

To His Highness the Viceroy.

" La Chénaie, *September* 28, 1857.

" At a sitting of the House of Commons, reference being incidentally made to the Suez Canal, Mr. Gladstone expressed himself in favour of the most recent project, and condemned the Government for opposing the manifest wish of the nation to participate in the execution of this enterprise. He said :—

" ' There is no one who, casting his eyes over the map of the globe, can deny that a canal through the Isthmus of Suez must be a great step towards the welfare of the whole world. This project commands the assent and sympathy of all the governments of Europe, especially that of France, our great ally.

Nothing, therefore, can be more deplorable than this conflict at Constantinople between the Ambassadors of France and England with respect to the canal.'

" *The Daily News*, in a leading article of the following day (September 10th), says :—

" 'This pretended right to keep the East for ourselves and exclude the rest of Europe from the Red Sea is the survival of an antiquated policy of which Lord Palmerston remembers far too much. This is a senile piece of nonsense on his lordship's part which ought to be got rid of for good, as it doubtless would have been if there had been twenty members present in the House who understood the question. For what have we to gain by excluding the European Powers from Asiatic waters ? France has aided us in our negotiations with Persia. Her co-operation is still more desirable in the war with China. Perhaps in the last century it might have been prudent and practicable to act alone in the affairs of the East, but at the present time there is no Power which does not stand in need of allies either in Europe or Asia. We need hardly point out that our best ally is France. The policy of the Cabinet, or rather that of Lord Palmerston, during the past year, has been to defy all Europe, France included, as regards the Suez Canal, and to declare, "The Red Sea is mine; you shall not enter it." '

" Referring to the transport of troops over the Isthmus of Suez, *The Daily News* of October 2, 1857, said:—

"'Thus the English Government admits that the Suez route is the best for communication with India, and after stubborn resistance, broken down by necessity, resolves to send by this route some of the troops which are being despatched to the relief of our gallant soldiers in India. Nothing could be a more complete avowal of the utility of M. de Lesseps's scheme; and this action of the Government is the implicit condemnation of Lord Palmerston and Lord Stratford de Redcliffe, who have hitherto opposed the scheme. It would seem as if Providence had set itself to inflict upon them the chastisement which they deserve, by making them, so to speak, responsible before public opinion for the difficulties which their country is experiencing in putting an end to the calamities which are so preying upon its interests, its affections, and its power. . . . Lord Palmerston and Lord Stratford de Redcliffe have not seen or foreseen anything of this. . . . Lulled by a false sense of security, they have yielded to their inclination for making themelves disagreeable to others.' "

Note for the Emperor Napoleon.

"Paris, *October* 20, 1857.

" The facility with which the Suez Canal can be made has been proved beyond all cavil by the International Commission of Engineers. The hearty and unflinching concurrence of the Viceroy and the free offer of capital ensure the success of the financial operation. The unanimous wish of the various nations, expressed with remarkable unanimity by the voice of the press or the deliberations of official bodies, has acquired for the enterprise the sympathy and support of their governments, and the conclusive resolutions passed at twenty meetings in the principal manufacturing and leading towns in England, together with the manifestations of the Councils-General and Chambers of Commerce in France, have testified to the harmony of the two allied nations, and have isolated the egotistical opposition which in vain attempted to create discord between them.

" This being so, it is now my duty, as holder of the concession for the work, to proceed to Constantinople and negotiate with respect to the Sultan's authorization, which was not, strictly speaking, necessary, according to the principle laid down by the British Embassy *à propos* of the railway from Alexandria to Suez, but which the Viceroy thought it right to solicit, in order to show his deference for his Suzerain, and to avoid giving any pretext to those who were ill-disposed for justifying their opposition. I may reckon upon being supported at Constantinople by the legations of Austria, Russia, Holland, Belgium, Prussia, Sweden, Denmark, the Hanseatic towns, Spain, Portugal, Sardinia, Tuscany, the Two Sicilies, Greece, and the United States.

" In order to maintain the universal character of the enterprise, I shall address myself to the representatives of these Powers, as well as to the French Embassy, should Lord Stratford de Redcliffe use his influence to hamper the liberty of the Divan.

"It may be that this influence will not be exerted now that Lord Palmerston has been compelled by the attitude of Parliament and public opinion to modify the violence of his original declarations, especially since the occurrence of the horrible events in India, which have shown that 'there is no security for the future if the Government does not take effective steps for bringing the mother country nearer to her Eastern colonies, and unless the first of these steps is to secure the piercing of the Isthmus of Suez.'

"I do not ask the Imperial Government to take any initiative, or to abandon the wise reserve which it has hitherto observed; but if during my negotiations at Constantinople I should have occasion, in my quality of a Frenchman and holder of the concession for an enterprise in which France is interested, to claim the intervention of the French Ambassador, as well as that of the representatives of other Powers, I hope that M. Thouvenel's protection would be accorded me, and that the Emperor will be pleased to instruct him to that effect."

To H.I.H. Prince Napoleon.

"Paris, *October* 12, 1857.

"In compliance with your kind suggestion, I have the honour to enclose you the note for the Emperor, explaining the present state of affairs relating to the Suez Canal. I trust you will say all you can in support of the request that instructions may be sent to M. Thouvenel. The following are those already sent to the representatives of Austria at Constantinople and Alexandria:—

"'By reason of the keen interest which the Austrian Government feels in. the enterprise of the Suez Canal, the demands made by the Viceroy of Egypt in this matter are to be supported as efficaciously as possible by the Austrian agents in the East, acting in harmony with the French diplomatic agents.'

"Upon the other hand, I am assured of the support of the United States Minister, as the Washington Government regards opposition to the opening of the maritime canal as an infringement upon the freedom of the seas."

To Count Th. de Lesseps, Paris.

"Paris, *November* 3, 1857.

"I have just seen Prince Napoleon, upon his return from Compiègne, and he assures me that the Emperor is very favourably disposed and sees no objection to my claiming the support of M. Thouvenel within the limits of my note of the 20th ult., which Count Walewski has had before him. I am both inclined and advised to act with prudence, and I shall be careful to avoid any cause of conflict.

"I am personally very grateful to the Emperor for what he said to Prince Napoleon about me. He made no secret of his hearty wish for the success of the undertaking."

To M. Barthélemy St. Hilaire, Paris.

" CONSTANTINOPLE, *December* 16, 1857.

" I yesterday made my first visit to Reschid Pasha, who was reappointed Grand Vizier a short time ago, and to other Ministers and functionaries, and the first dragoman to the Embassy, who accompanied me, informed them all that he was instructed by M. Thouvenel how much interest his Government attached to the success of my negotiations with them.

" Reschid Pasha seemed very pleased at this resumption of relations with the French Embassy, and in two or three days Aali Pasha, the Minister of Foreign Affairs, will give a grand dinner, to which M. Thouvenel, Reschid Pasha, and myself will be invited.

" Reschid knows perfectly well that the French Embassy is going to give his temper and disposition a fresh trial, and he is too anxious to remain in office to compromise himself if he can help it. I shall not commence my parleys with him and the other ministers until after this dinner of reconciliation. However, I am not losing any time, and am preparing my ground in all directions, for there is in all countries, even in Turkey, a public opinion of which account must be taken, and in neglecting no opportunity, great or small, of obtaining partizans, I help the work on."

To Count Th. de Lesseps.

"Constantinople, *December* 25, 1857.

"Yesterday I had a conference, extending over two hours, with Reschid Pasha in his house at Emerghian, on the Bosphorus. I did not fail to say all I could think of as likely to strike him, and show him the advantage of a favourable solution emanating from the initial action of Turkey herself.

"Reschid brought me back in his steamer, and as we were alone we were able to carry on the conversation. He readily made me formal promises, and I was even astonished to find how very strongly he expressed himself in favour of the canal.

"I gave him to understand that I set less store by his promises than by the manner in which he carried them out, either upon his own responsibility, or at the orders of the Sultan or the Cabinet, in the event of his not caring to take the personal responsibility of the matter. I learnt that upon leaving me he lost no time in submitting to the Ministerial Council a memorandum which I had previously shown to M. Thouvenel and of which he expressed his approval. I send you a copy of it for Count Walewski. Previous to my conference with Reschid I had a separate interview with each member of the Council, and I did all I could to win their ear in favour of the enterprise. I have also had one or two important conversations with Nedgib Pasha, whom the Sultan had recently sent to Egypt. He is a sort of steward of the Harem, and he is in such favour with his sovereign that the ministers have to keep on good terms with him.

" My arrival at Constantinople was very opportune, as the intrigues of the English Embassy, which have been at work for the last three years, were beginning to tell, and threatened to take root.

"You can tell the minister that M. Thouvenel never goes too far, and is not at all likely to compromise himself; but few ambassadors could do what he can in a country of this kind, so long as he is left free to act in his own way. The representatives of the foreign powers continue to aid me with their advice and influence, and I have communicated my memorandum to each of them. *The Times* correspondent is sending it to his journal.

" I have now something confidential to tell you which will explain why Lord Stratford de Redcliffe went on leave before my arrival. I learn from a foreign source that during the visit of the Emperor and Empress to Queen Victoria, at Osborne, the Suez Canal question was discussed at a conference attended by Lord Palmerston and Count Walewski. As the Prime Minister could not get the French Government to use its influence here against the canal, the only thing done was to renew the agreement that the diplomatic agents of both countries should remain neutral in the matter. This was equivalent to admitting that the neutrality had been violated, as indeed Lord Palmerston had already declared in public. In any case, this principle is again to be adopted in theory, but if in practice we are weak enough to carry it out I am ready to prove now that the English will not. In order to have the appearance of doing so, they have sent Lord Stratford de Redcliffe on leave and put in his place Mr. Alison, his first secretary, who is not less devoted than himself to the Foreign Office, while in Egypt the honest and trusty Mr. Bruce is replaced at the Consulate-General of Alexandria by Mr. Green.

" Count Walewski, who was present at the Osborne conference, will be able to tell you whether I am right."

Memorandum to Reschid Pasha.

" CONSTANTINOPLE, *December* 29, 1857.

" I have the honour to request your Highness to apply to the Sultan for an Iradé authorizing the Commercial Company, of which I am the representative, to execute the works intended to effect a junction between the Mediterranean and the Red Sea by means of a maritime canal.

" At the time of my first visit, three years ago, to Constantinople, during which your Highness was kept duly supplied with all the preliminary documents, you were pleased to write me a letter (March, 1855) in which you spoke of the enterprise as being 'most useful,' adding, 'in conformity with the Imperial order relating to this interesting undertaking, the question is now before the Cabinet Council.'

"Since then, in order to facilitate the examination and decision of the Sublime Porte, I have endeavoured to clear away the objections urged as to the possibility of the enterprise, or the fear of its being inimical to the legitimate interests of foreign powers. The first objection has been disposed of by the report of the International Commission of Engineers, and the second by the unanimous expression of public opinion in all countries. The adhesion of the Continental governments has been not less explicit, and with regard to England I think it well to mention the last official statements made in the House of Commons on August 14th ult., subsequent to the resolutions adopted by the Associations and Chambers of Commerce, and by the many meetings held in the principal towns of Great Britain.

"At this sitting of the House, Mr. Gladstone expressed himself as follows:—

"'The House ought to treat the Suez Canal scheme, as well as the Euphrates Railway and the telegraph schemes, as a purely commercial question, acting upon the assured principle that the best judges of a commercial speculation are those who have undertaken to put capital into it. If this question should ever be converted by the Government into a political one, there would be every danger of a break in that European concert and agreement which are of such capital importance as regards our Oriental policy. Yet no one can look at a map of the world and deny that a canal through the Isthmus of Suez would, if it were practicable, be of great service to humanity. This project has been approved and found excellent by all the governments of Europe, especially by France, our great ally. What could be more unfortunate, therefore, than to find quarrels arising on this subject between the ambassadors of our two countries at Constantinople? Bearing in mind our Indian possessions, do not let us give room in Europe for the belief that, for the maintenance of our rule in India, it is necessary that we should oppose measures which are advantageous to the general interests of Europe. Do not let us allow so deplorable an inconsistency to take root, for this would weaken our power in Hindostan more than ten such mutinies as that which has just occurred.'

" Lord Palmerston replied :—

" ' The chief and only motive that we have urged upon the Turkish Government against accepting the proposed plan is not the injury caused to England, but the injury caused to Turkey, the danger of impairing the integrity of the Ottoman empire. '

" The whole question, therefore, is now confined to a right understanding as to what the interests of the Ottoman Empire really are. It is clear that this can only be known to the Government of the Sultan, to which I appeal with the conviction that the careful examination which it has already made will have demonstrated to it the many advantages which Turkey must derive from the execution of the Suez Canal. In explanation of this it is only necessary to remind you that the route from Constantinople to the Indian Ocean will be abridged by 4,800 leagues, that the Ottoman possessions of Arabia and the East Coast of Africa will be brought within touch of the metropolis, and that the easy access to the Red Sea will be an inestimable advantage for the Mussulman pilgrims to the holy places.

"When the Imperial Government has given the opinion which it deems suitable to its interests, it will also be free to declare that the maritime canal is to be open at all times as a neutral passage to all the merchant vessels going from sea to sea, without any exclusive destination, or any preference as regards nationality. The accession of the foreign Powers, whom the Sublime Porte will doubtless invite to give their adhesion to its declarations, will be no more than the outcome of a fact which the Porte has already decided to accomplish in keeping with its competency and rights. This was the opinion expressed by Prince Metternich in the course of an interview which I had with him, and which was communicated by me to the different cabinets in Europe and the United States, whose representatives at Constantinople have received instructions to support my action.'

"These considerations will form the elements of our negotiations, and I am at your Highness's disposal and at that of the Sublime Porte for any further information or explanations which may be deemed necessary. I am convinced that at a moment when the most enlightened men in the Ottoman Empire are happily united in order to carry out the liberal intentions of their sovereign, the project for piercing the Isthmus of Suez will, after having been consecrated by science and public opinion, meet with a favourable reception from the councillors of the Sultan."

To Count Th. de Lesseps, Paris.

"CONSTANTINOPLE, *January* 11, 1858.

"Here is an unfortunate occurrence which will probably have an awkward effect upon the negotiations relating to the canal. I refer to the sudden and unexpected death of Reschid Pasha. I had seen him the day before, and he was in excellent health. I am told that after drinking a cup of coffee he was seized with convulsions and vomiting, and soon expired. In order to put an end to all the rumours in circulation, a commission of European physicians was appointed, and though they were unable to make a *post-mortem* examination, they issued a report that the death was due to natural causes. The people of the East are very slow to believe this when a great personage disappears. Be this as it may be, I regret his death in a double sense : in the first place, because it is a personal loss ; and, in the second place, because he seemed to have shaken himself pretty free of English influence in regard to the canal.

"His successor, Aali Pasha, is beyond all question the most upright and best informed man in the Empire, but he is extremely timid, and reluctant to take any initiative. The threats of Lord Palmerston after the Congress of Paris will always be ringing in his ears. In any case, I shall be on the best of terms with him personally, and he will have the wish, if he has not, as I fear, the power, to keep his promises."

To M. Thouvenel, Constantinople.

"CONSTANTINOPLE, *February* 6, 1858.

"I had a long conversation this morning with Aali Pasha, and explained to him our mutual situation with the utmost frankness, and communicated to him the reports which I had received from Paris, London, and Egypt. Finding that I did not wish to press him too closely, and that I took into account the difficult position in which he was placed, he made no secret of the fact that he was desirous of awaiting the result of the questions which were going to be put in the House of Commons. I handed him the extract of the instructions which the Viceroy of

Egypt had sent me, and he expressed his hearty concurrence in the friendly sentiments which Mohammed
expressed.   I also read him the following letter,
which I had received from Cairo under date of February 6th :—

"'The day before yesterday the English Consul, Mr.
Green, went to see the Viceroy and read him a letter
from Lord Clarendon, thanking him on behalf of the
British Government for the facilities afforded in the
transport of troops to India. But he added that none
of the news sent by M. de Lesseps with regard to the
progress being made at Constantinople in carrying
the canal scheme through was in keeping with his
information; that Mr. Alison, the English Chargé
d'Affaires in the absence of Lord Stratford de Redcliffe, had shown Aali Pasha letters from Lord
Palmerston in opposition to the canal, and that Aali
had signed an agreement not to grant the firman
without the assent of England. These details were
repeated almost publicly in front of the Viceroy's
palace, in the presence of several persons, by Mus-
tapha Bey, the Viceroy's nephew. The Viceroy is
said to have very sensibly replied that, so far as he
was concerned, he had granted the concession for the
canal three years ago ; that he was no longer in a
position to interfere ; that the matter rested with the
Divan ; and that if England had anything to say she
must address herself to the Porte.'"

To Aali Pasha, Grand Vizier.

"Constantinople, *February* 24, 1858.

" As it may be useful that you should know the impression of foreigners, especially of Englishmen, as to the Suez Canal, I think it well to communicate to you the contents of a letter which I have received from an Englishman in London. Many of the remarks made in this letter, which I will ask you to return me, are full of common sense, frankness, and verity.

"It is, in truth, quite time for Turkey, in the interests of her own dignity, to come to a decision. I quite understood, as I told you yesterday, that circumstances would not admit of your keeping the promise which you made of obtaining this decision by March 3rd ; but allow me to remind you that it will be difficult, if not impossible, for me, in view of the instructions which I showed you, to wait beyond the 15th of that month. It scarcely seems to me that the colds from which several of your colleagues are suffering will be a sufficient reason for adjourning this matter, which has been under consideration for three years ; and it is one in which the Grand Vizier alone is responsible for the decision, right or wrong, which may be come to. I wrote yesterday to the Viceroy to inform him that your Highness distinctly denied having allowed any foreign Power to fetter your liberty of action, and that you had made no declaration, either *verbal* or *written*, to any foreign diplomatist."

The following is an extract from the London letter
and the article which accompanied it :—

" Letter :

" ' Are they so blind at Constantinople as not to
see that they are making over their dependency to
England, who deceives them, frightens them, and
consequently despises them. If the Sultan acts
according to his own responsibility, England will
respect him in consequence, but will never do him
any injury. When will the Turks wake up and issue
the firman ? They have been asleep long enough.'

" Article :

" ' Under Lord Palmerston's Ministry threats were
addressed, both in London and at Constantinople,
to the higher agents and functionaries of Turkey.
They were told that if the Porte showed itself favourable to this enterprise, it would earn for itself the
lasting hostility of England, and that, in addition, it
would probably bring about a struggle between
France and England by which Turkey would be the
sufferer.

" ' When Lord Derby succeeded Lord Palmerston in
office, the Porte thought to avail itself of the change
to grant the firman which the Viceroy had asked for,
and a telegram was sent to Musurus, the Turkish
Ambassador in London, requesting him to inform
Lord Malmesbury (who was Secretary of State for
Foreign Affairs in Lord Derby's Ministry) that the
Government of the Sultan, not wishing to take any

action in this matter *unknown* to the British Government, would be glad to know his views on the subject. Lord Malmesbury replied that he and his colleagues shared the opinion of the previous Ministry, and that they should continue to oppose the undertaking. He added that he noticed with pleasure, in the communication from the Porte, that nothing would be done without *the consent of England.* This was how he was pleased to translate the word "unknown" (*insu*).

"'This reply excited considerable astonishment at Constantinople. The Divan lost no time in instructing Musurus to declare that they had never dreamed of alienating their liberty of action in a question of internal administration, or of making their decision dependent upon the *fiat* of a foreign government; and, finally, that if, out of deference for an ally, they had announced their intention of taking no action *unknown* to that ally, they had no idea of allowing the solution to depend upon the *consent* of the British Government.

"'Such, from the diplomatic point of view, is the present condition of affairs between the two governments. What you may regard as quite certain is that the Turkish statesmen, finding that the English Cabinet does not dare to admit openly in Parliament the steps taken by its diplomatic agents, sees how puerile and useless is an opposition which cannot face a public debate.'"

To Count Th. de Lesseps, Paris.

"Constantinople, *March* 30, 1858.

" M. Thouvenel has written me this morning as under :—

" 'I have seen Aali Pasha and Fuad Pasha, and I find them both of the same way of thinking that they were,—viz., very favourable to the canal, and anxious to make it clear to the world at large that the Porte does not of itself raise any difficulty in the way of your great enterprise. Aali spoke in a firm and decisive tone which augurs. well for us, and he was very pleased at what I had to tell him.' "

To Mr. D. A. Lange, Agent of the Suez Canal Company in England.

"Constantinople, *April* 15, 1858.

" I conveyed to you briefly, in my telegram of the 11th inst., my views as to the reply made by Mr. Disraeli in the House of Commons. This telegram was as follows: 'Mr. Disraeli talks of the sanction of England. Such a pretension is absurd. No one wants any sanction of the sort. The only question is, does Lord Derby intend to go on threatening Turkey, which wishes the canal to be made, as Lord Palmerston did?' "

"I hope if fresh questions are put and a debate follows, no more such disingenuous side issues will be raised, for they do not redound to the credit of your Parliament. But the success of our enterprise cannot be compromised by an opposition of this kind, and the affair, I am thankful to say, has now reached a point which makes us independent of the antiquated policy of some of your statesmen.

"I have attained a situation which, thanks to the forces placed at my disposal, enable me to withstand the efforts of my opponents. I will repeat here what I said last year at a meeting in London, my remarks commanding unanimous assent: 'My enterprise will not be carried out by those who are against, but by those who are for it,' and as the latter are more numerous and stronger than the former, and as, moreover, they are in the right, I shall take the liberty of going forward, and of taking practical action, doing without those who stand in my way.

"The Porte, which *stoutly repudiates any common share in the opposition of the English Government,* is awaiting the public explanations which Lord Derby's Ministry promised to make touching its policy in this matter. If these explanations are ambiguous, or if they are openly hostile, the Canal Company, armed with the Egyptian concession, to which the Sultan cannot offer any opposition, will take its own course and enter upon the work with the capital which it has at its disposal. This will be the most effectual mode of replying to the ceaseless objection that the enterprise is impracticable."

To Count Th. de Lesseps, Paris.

"Constantinople, *April* 17, 1858.

" The intentional dodging and backing-out of the question which characterises the action of the English Cabinet are not likely to stop much less to turn me back. I look upon them merely as so many posts which I leave behind me as I go on, and which will soon serve to measure the distance which I have travelled over.

" I do not understand why some politicians, whose advice I generally follow, regret that I am here instead of in London. They will not adhere to this view if they will compare the present position of my enterprise with what it was when I left Paris four months ago. Lord Palmerston had then publicly stated that the question was one for Turkey, not for England. It became necessary, therefore, to cut off the retreat on Constantinople which he was keeping open for me, and from which he would have beaten me, for, with the threats of the English agents and the passive attitude of ours, I am now more than ever convinced that but for my presence here Lord Palmerston or his successors *ejusdem farinæ* would have wormed out of the Porte some declaration fatal to the making of the canal.

" Such a stratagem has now become impossible, because I am able to keep a look-out for, and to ward off, the blows aimed at us. At the present time,

Turkey unequivocally repudiates any solidarity with the English opposition, and this it is which constitutes my strength and will enable me to go forward unmoved towards my end, whatever may be the re sult of the explanations Mr. Disraeli is to offer in the House of Commons.

" The Porte has promised to send this very day a telegram to M. Musurus, instructing him to inform the English Cabinet that it repudiates all solidarity in this opposition to the Suez Canal."

To M. Barthélemy St. Hilaire, Paris.

"CONSTANTINOPLE, *April* 22, 1858.

" When we have made it clear that Lord Derby's Cabinet has succeeded in eluding or in burking a parliamentary debate upon the Suez Canal, or if the Government makes a positive declaration of hostility to it, we shall be in a position to send to all our agents and correspondents the following memorandum, which please submit to my brother for translation, telling him that I will keep it back if necessary, but that my mind is made up. Lord Palmerston, by throwing the responsibility for the opposition on Turkey, had already contributed to advance the question of making the Suez Canal.

" ' Mr. Disraeli's declarations have decided the execution of our great work. Statesmen who represent neither the ideas of their age or country have dared to publicly denounce as chimerical a project elaborately worked out by very competent engineers, agreed to by the principal men of science in Europe, and accepted by all the great commercial and maritime towns, including those in England.

" ' As there is no more serious resistance than this to be encountered, and as English statesmen have no better reasons than these to justify the hostile action of their agents at Constantinople for the last three years, all that we have to do is to prove that the so-called chimera is a reality.

" ' The Universal Company of the Suez Canal, armed with the regular concession of the Viceroy, to which the Sultan offers no opposition in so far as concerns his Suzerainty and the interests of his Empire, will, however, be too prudent to provoke a conflict between the policy of progress and that of retrogression, or to give its opponents an excuse for playing upon prejudices; while so as to avoid all misunderstandings in an affair which should retain its general and commercial character, the Company will not ask for the assistance of any of the governments of whose support it was assured. But it is about to organize itself in a definite form; it will march resolutely forward and complete its work, backed up by the investments of its shareholders of all nations, and by the public opinion of the whole world.

" ' The Scientific Commission will meet about the end of June, and its report will settle the conditions under which the works are to be executed, in order to open the first section of the canal. A temporary board of administration will then decide how much capital is to be issued; the shareholders will receive intimation of when they are to pay their calls, and every arrangement will be made, so that by the end of the year the work may be put thoroughly into hand, and carried on without interruption.'

"I sent Aali Pasha a letter of the 15th, containing a copy of your capital answer to Mr. Disraeli. I conferred with him yesterday, and read him a copy of the above circular. He quite understands that I have no other course open to me, and he prefers that I should admit that Turkey does not oppose our enterprise, so far as concerns her interests, than that I should be constrained to record the fact that she submits, and without any counteracting good, to foreign pressure.

"We are, therefore, quite agreed, and I am glad, taking everything into account, that I resolved not to ask, for the present, of Turkey more than she can, as she is situated, well agree to.

"It is no use deceiving oneself as to the situation, which I think that I can see very clearly.

"When it is a question of despoiling others for the common benefit, the English give each other a hint, and leave the Government to do as it pleases. So they will be banded together against us in this business. Continental governments, which often struggle with one another upon questions of existence—a situation of which an island power like Great Britain profits largely—will not care to create external embarrassments for themselves, and will not hold out a helping hand to us if we are unable to get along by ourselves. I have shown you how things stand with Turkey. Egypt has done all she can be reasonably asked to do. She is not in a position to support alone any longer the responsibilities of the enterprise. Admitting that the Viceroy were disposed to do so, I should not advise him to take such a responsibility on himself. The incessant intrigues of the English agents would eventually kill him, or, with his nervous and irritable temperament, would drive him out of his mind. The course which I have decided upon is therefore the only one possible, and we must gather up all our energy, and that of our friends, in order to march on to the goal, and not to allow ourselves to be deterred from our course.

"The English policy has been to have a double shot, by seizing Perim and opposing the canal. If the policy of the Western Powers and of Turkey is powerless as concerns Perim, our company is not going to haul down its flag. It will be stronger than Lord Palmerston and Mr. Disraeli.

"What I have decided upon will be carried out by the end of the year, except in the improbable event of Lord Derby's Cabinet declaring *explicitly* that England renounces all opposition and leaves Turkey full liberty of action."

To M. de Negrelli, Vienna.

"Constantinople, *April* 24, 1858.

"I have received yours of the 14th, and I have read as usual with extreme care the particulars of your recent conversation with Prince Metternich, whose great ability and rectitude of judgment are unimpaired. He is quite right; our enterprise is ripe, and we must not be any more disheartened by what Mr. Disraeli says than we were by the utterances of Lord Palmerston; while we must, at the same time, calmly consider the position in its true light, without being too sanguine, but also without hesitating or taking a single step backwards.

"You will see by my enclosures that I have acted in accordance with these precepts, and have taken the only course which in the circumstances was open to me. This being so, it would be imprudent to thrust France into the foreground. To do so would be almost an act of political antagonism.

"Our affair is, to my mind, in the best possible position. My agreement with the Turks enabled me to go steadily forward, and you will see that eventually every one will follow in our wake when it is found that we are not to be intimidated.

"When the time arrives for securing subscriptions we shall be overwhelmed with applications, whatever may be the case with other financial operations. In France, the opposition of England will be the chief

source of attraction for us.   You may rely on me that
this will be so.   You know that I am not prone to
exaggeration, but as all the information relating to
this undertaking is centred in me, I know that we
have even more power than the most sanguine of my
friends can imagine."

To M. Thouvenel, Constantinople.

"CONSTANTINOPLE, *April* 28, 1858.

" I received, last night, the following telegram :—
" 'Questions will shortly be asked by Roebuck.   It
will then be seen that, despite the tactics of Lord
Malmesbury and *The Times*, England wishes the canal
to be made.   Try and come.' "

To Aali Pasha, Grand Vizier.

"CONSTANTINOPLE, *April* 28, 1858.

" I beg to forward to your Highness the original of
the telegram I have just received from London, and I
also enclose, in order to keep you posted in the action
I am taking, copies of a letter to M. de Negrelli, at
Vienna, and of the instructions sent by me to M.
Barthélemy St. Hilaire."

To M. Barthélemy St. Hilaire, Paris.

"CONSTANTINOPLE, *May* 12, 1858.

" My object, which was that of showing that I am
resolved to go on in spite of all opposition, having
been attained here by the communication of my pro-

posed circular, and in England by my letter to Mr. Lange, we can now await the first discussion which is to be raised in the House of Commons.

"This debate, and the resolution which is to be moved in connection with it in the beginning of June, as Mr. Lange writes me, are, moreover, facts which must modify my plans. Instead of remaining here until after the debate, I have determined to go to England, taking Paris on the way, and then to return here previous to constituting the company, with or without the Sultan's ratification. I have just advised the Viceroy of my intentions."

To M. de Negrelli, Vienna.

"ATHENS, *May* 21, 1858.

"As I had advised you was my intention, I sailed on the 19th for Marseilles, and shall probably be in London by the end of the month. I will telegraph you what is done. I allow our great undertaking to be guided by the course of events, and when the wind changes it is necessary for one to shift one's sails, heading as much as possible for one's destination.

"I expect to be back in Constantinople in a month. If the English Ministry renews its declaration of hostility in Parliament, I shall be obliged to agitate anew in England and elsewhere. I shall publicly announce my intention of forming the company, in conformity with the plan which I described to you after the previous declarations made by Mr. Disraeli."

To M. Barthélemy St. Hilaire, Paris.

"PORT FIGARI (SARDINIA), *May* 27, 1858.

"As we were leaving the Straits of Messina, a storm and the breaking our screw placed us in a state of peril for three days, and we should probably have been driven ashore but for a small steamer which carries the mails between Genoa and Cagliari, and which, despite the heavy seas, pluckily came to tow us in here, where we arrived after great difficulty. This accident will prevent my reaching London in time to be present at the debate. Tell Lange that if our supporters fail to stop the opposition of the Government, it will be powerless to impede the progress of a private enterprise, and will create everywhere a very bad feeling against England."

To M. de Negrelli, Vienna; M. Ruyssenaers, Alexandria; and to M. Charles Aimé de Lesseps, Constantinople.

(By Telegraph.)

"LONDON, *June* 8, 1858.

"The debate in Parliament, which made an impression very favourable for us upon public opinion, will be followed by fresh motions. The Ministry will be beset with questions until the end of the session. The ability and persistent energy of our partizans ensure a moral success. The general opinion is that the onward progress of the company cannot be arrested, and that the opposition will be unable to hold its own. I shall very shortly return to Egypt and Constantinople."

To M. Barthélemy St. Hilaire, Paris.

"LONDON, *June* 9, 1858.

" I send you the substance of my conversation with our ambassador, the Duc de Malakoff :—

" 1st. The marshal is very well disposed towards our enterprise.

" 2nd. He has no instructions to take any action here.

" 3rd. He seemed relieved when I told him that I had come to London upon business relating to the canal, and had no need to ask for his intervention.

" 4th. My assurance and the declaration which I made him of my intention of following up the enterprise and carrying it into *execution,* despite the opposition of the English Government, created a very favourable impression upon his mind, and upon his attitude towards me, as he expressed the hope that he might live to assist at the inauguration of the canal.

" At a large dinner and evening party given by Mr. Hankey, M.P., governor of the Bank of England, several members of the house who were formerly opposed to the scheme, assured me that I had *converted* them.

" In short, after having heard many opinions, I judge the situation in England to be pretty much this :

" The sixty-two members who voted for Mr. Roe-

I 2

buck's motion have quite made up their minds on the
subject, and will always vote in our favour.   The
remainder, who form the docile ministerial majority,
have reserved their opinion with regard to the canal,
at the request of Mr. Disraeli, in order to gain time to
acquaint themselves with its merits before voting for
or against it.   A large proportion of this majority is,
according to what Mr. Roebuck himself told me, sys-
tematically hostile to the canal, because it is *syste-*
matically hostile to France.   Lord John Russell, Mr.
Milner Gibson, Mr. Roebuck, and others are going to
come to an arrangement for enlightening the House
by means of fresh resolutions, so as to force the
Ministry in their stronghold.   The following is the
telegram which I have sent to Vienna, Alexandria,
and Constantinople, defining our position :—

" ' With regard to the communications of France
and England concerning the canal, it had been agreed
in principle that, in view of the fact that the two
governments held different opinions, the enterprise
should be allowed to take its own course, the more so
as it did not demand the assistance of any government.   The French and English diplomatic agents at
Constantinople and Alexandria were to remain neutral, and abstain from bringing their influence to bear.'

" It is, then, most dishonest to assert that France
does not take any interest in the canal, because the
French agents have been true to the principle of
neutrality which they were instructed to observe, and

because the English have been untrue to it both in Turkey and Egypt."

To Mr. D. A. Lange, London.

" Corfu, June 28, 1858.

"The communications I made to you in London demolished the arguments of our adversaries as to the *alleged* indifference of the French Government; the latest revelations made at Constantinople prove that the second assertion as to the opposition of the Porte is equally false, and that it is the English Government, the representative of a loyal, powerful, and civilised people, which has not scrupled to employ the arms of the weak and the barbarous—that is to say, hypocrisy and cunning—and to conceal its opposition behind a door (*porte*) which it thinks it can open and shut as it pleases.

" I may now proceed to dispose of the third assertion, touching the connivance of Austrian diplomacy with the hostile manœuvres of the British Cabinet.

" I saw, while passing through Vienna, several of the Emperor's ministers and various personages who told me how things stood in Austria. I shall be glad if you will communicate the information to our friends in the House, but do not make it public.

" It is evident that the House of Commons was led astray in the debate of June 1st, not only by Stephenson, but by the utterances of ministers. The majority, obtained by underhand intrigue, despite the

admirable speeches of the minority, must not lead the
English Ministry to suppose that it can continue
practices at Constantinople which I am determined
most resolutely to withstand, and which, if they were
resumed next month during my negotiations with the
Porte, might lead to a deplorable conflict.

"I beg of you expressly to let your fellow-countrymen clearly understand that I am not to be blamed
for any such conflict should it arrive, and that I have
forewarned all my English friends of the many embarrassments which the absurd and unbearable policy of
their Government in this matter of the Suez Canal
would probably bring upon their country.

"While showing every readiness to go on with the
negotiations, I am making my preparations to get the
company in working order, and commence operations
before the end of the year.

To M. Thouvenel, Constantinople.
"Corfu, *June* 28, 1858.

"While on my way here I met Fuad Pasha, who
saw the Emperor during his stay in Paris. He could
not forget the wholesome rebuke of the Emperor about
'a firman relating to Egypt,' and he asked Count
Walewski what this rebuke meant, but our minister
declined to give him any explanation. I thought it
my duty to tell him as a *friend*, and as one holding no
official position, that if the Emperor was vexed it was
doubtless because he thought that in a matter of this

kind Turkey ought to have displayed more initiative and vitality, instead of attempting, as she seems to have done, to *create a political question between France and England.* For upon what ground does the Porte consider itself bound to consult England about the Suez Canal, when she did not consult France with regard to the concession of the Euphrates Railway? One may, without being unduly susceptible, resent this conduct of having two weights and measures.

"I have received the following letter from London, under date of June 22nd :—

"'You are strongly advised not to delay commencing the execution of your work. Your course will be watched with the keenest interest by the members who wish you *every success*, and even by some of those who voted against Mr. Roebuck's motion. It is the universal opinion that the political question will be settled by the common sense of the English people, and you may be sure that as soon as it is seen that you are about to begin making the canal there will be a very great change.'"

To M. Barthélemy St. Hilaire, Paris.

"ALEXANDRIA, *July* 9, 1858.

"I was with the Viceroy when the news of the terrible massacre at Jeddah arrived. Upon my expressing my indignation, he quietly observed : 'What! you, who have known the East so much longer than I, are surprised. But your experience ought to have

told you that when fanatical and barbarous populations
are not kept tightly in hand they are certain, one day
or another, to indulge in the most deplorable excesses.
Even here there are many people who greet you with
respect who would tear your heart out if they dared.
English policy wrested the administration of Syria
from my father, and there will be other examples of
what unbridled fanaticism is capable of. But as to
Jeddah and Arabia, *our canal* will put a stop to all
that, and Arabia will inevitably be brought into line
with Europe.'

"These very pertinent observations are worth
recording.

"It may be of interest to give you some particulars
about what occurred at Jeddah. I have them from
Mdlle. Elise Eveillard and from M. Emerat, who escaped
from the massacre, though they were very severely
maltreated, and are still suffering from their wounds.

"Five thousand rioters swooped down upon the
French and English consulates. The English consul
was literally cut to pieces, while two of his dragomans
and an Indian servant had their throats cut. The
French consul, M. Eveillard, was stabbed and hacked
to death; his wife was killed by a stab in the breast,
after having killed one native and wounded another.
His daughter, while this terrible scene was being
enacted, had her father's head, cut open by two sabre-
strokes, resting against her knees; and seeing M.
Emerat, the chancellor of the consulate, who had

already received three wounds, engaged in a hand-to-hand struggle with one of the rioters, she had the courage to make a spring at him, bury her nails in his face, and bite him in the arm until he dropped his weapon, which M. Emerat was then able to pick up and use against fresh assailants, until at last he fell exhausted and bleeding. Mdlle. Eveillard had her cheek cut open by a yatagan, and had sunk to the ground. The assailants, thinking that they were both despatched, proceeded to pillage the house, and Mdlle. Eveillard covered herself and the bodies of her parents with the cushions of the divan in the hope that they would all escape notice. Soon after a fresh band of rioters came into the room, and seeing legs emerging from the coverings of the divan, gave several sword-thrusts at them to see if the bodies to which they belonged were really lifeless. Mdlle. Eveillard had the fortitude to make no movement, and the men went away. But even then her sufferings were not over, for the men came back, and in order to see if a large cupboard, at the foot of which she was lying under the cushions, contained any valuables, four or five of these wretches stood upon them. It may be imagined what her agony of body and mind must have been. At length this band of savages, drunk with blood and pillage, made off.

" There then arrived a young negro, who had been sent to her rescue by the ladies of an adjoining harem to whom Madame Eveillard and her daughter had a few days before taken some medicine. This young negro, alone amid so many bloodthirsty enemies, had been obliged to play a passive part until the sun had gone down, when he made Mdlle. Eveillard understand by signs that he had come as a friend. He rescued her from the living tomb in which she lay, and after many hairbreadth escapes brought her in safety to the harem, where she was very hospitably treated.

"M. Emerat had been rescued by an Algerian Mussulman who had served for twelve years in the French army, and who had fallen upon the rioters with great pluck when he saw them cut down the consular flagstaff and trample the tricolour under foot. He succeeded in conveying M. Emerat to a place of safety."

To M. Ruyssenaers, Alexandria.

"Constantinople, *July* 28, 1858.

"We have every reason to be satisfied, for I have just raised the curtain upon our last act. It was no use wasting precious moments with the Turks, but, taking advantage of their declarations, I have put on record the fact of their tacit adhesion, and have placed my interests and those of the company under the irrefragable protection of the Emperor of the French.

"Baron de Prokesch, ambassador of Austria; M. de Boutenieff, ambassador of Russia; General von Wildenbruck, minister of Prussia; Señor de Souza, minister of Spain, and the other diplomatic representatives at Constantinople approve of my determination, will inform their respective governments of it, and will if necessary co-operate with the ambassador of France.

"Please inform the Viceroy of what I have done."

To M. Thouvenel, Constantinople.

"CONSTANTINOPLE, *July* 30, 1858.

"The conversation which I had with Aali Pasha on my arrival convinced me that, owing to the continuous action of the English Embassy, as well as to the discussion in the English Parliament on June 1st, the Sublime Porte is so situated that it feels the necessity of having a counterpoise which would enable it, without exposing itself to formidable difficulties, to go through the official formality of according a sanction which it has already given in principle. It undoubtedly exaggerated these difficulties, for had it followed its own inspirations it would not have created for itself more embarrassments than its vassal the Viceroy, whose conduct in this matter has won him universal sympathy, has had to face.

"But you know better than anyone how Turkey is situated, and will therefore understand her passive attitude in the matter.

"As the ministers of the Porte had often declared to you that they were favourable to the canal scheme, and that their government did not raise any difficulty in the way of its realisation *proprio motu*, it seems to me that there were no further negotiations to be pursued with them. I then discussed the state of affairs with Sir Henry Bulwer, whom I had formerly known personally well enough to admit of my explaining my views to him with regard to the false and equivocal position in which his government in my eyes placed itself. The English Embassy, I said, had hitherto shown itself very hostile to my enterprise, and yet had not taken any official or ostensible step to justify its opposition upon the ground of English interests being imperilled. Mr. Disraeli's utterances in the debate of June 1st are a proof that what I say is true.

" I have informed you of my conversation with Sir Henry Bulwer, and I now send you a copy of the letter which, at his request, I wrote to him on the 28th inst., and in acknowledging its receipt he tells me that he is about to transmit it, together with the documents I sent him, to his government, and will await their instructions. It will therefore be for the Imperial Government to protect my rights and those of the company.

" I shall continue, for my part, to do all that lies in my power to secure the aid, if we shall require it, of the other governments from which I have received the most favourable assurances of good will."

To M. Barthélemy St. Hilaire, Paris.

"CONSTANTINOPLE, *August* 18, 1858.

" I have just made arrangements at Odessa for appointing agents of the canal company in Russia. I have advised all the foreign embassies of my departure on the 21st for the purpose of constituting the company, and I have sent them copies of my letters of the 28th and 30th ult. to Sir H. Bulwer and M. Thouvenel. As the political question with regard to England has been left to our government, and as the tacit adhesion of the Porte has been made sufficiently clear, there is no reason for delaying any further the organisation of the company.

" M. Thouvenel approves of my plans, and sees no further need for me to remain here; for, as I have pointed out to him, if I awaited here the decision of the Imperial Government, I should be obliged to submit to the delay which is certain to occur in the negotiations between Paris and London, whereas I am anxious to get our board of directors together.

" Mr. Stephenson admits, in a letter to *The Times*, that he only visited a part of the isthmus. I know what part that is, for I myself saw the tracks of his carriage wheels, which did not extend more than a league beyond Suez. He omitted the most essential part of the excursion—viz., to the Bitter Lakes, from Lake Timsah to Pelusium and the Mediterranean coast, for that is where the only difficulties were to be met with, difficulties which ill-will and ignorance have been pleased to exaggerate. With regard to the substance of his letter, it seems to me to contain only bare statements, without any argumentative reply to the reports of the International Commission, or to the scientific observations of Messrs. Paleacapa, de Negrelli, Conrad, and Dupin, the reporter of the commission of the Académie des Sciences."

To M. de Negrelli, Vienna.

"PARIS, *September* 14, 1858.

"Since my return here I have been devoting my whole time to the establishment of agencies for the company abroad and in France, as well as of scheduling the private subscriptions which have been sent to me, and which already reach £3,200,000. The adversaries of the enterprise, our faithful allies over the water, have already lost their two first campaigns as to the impossibility of making the canal and the hostilities of the Porte. All their efforts are now directed to deterring their compatriots from subscribing to it, because, in their innate pride and insular ignorance, they believe that their example will prevent other nations from investing money in it. We are now in course of destroying their last illusions.

"The Emperor is in favour of subordinating the political question to the organisation of the company, which will be strong enough to withstand opposition, and which the Continental governments will be in a position to support if needful. This seems to me very prudent, and is quite in keeping with my view as to government intervention, which should follow if the necessity for it arises, and not precede the execution of a commercial and industrial enterprise.

"The main thing is that I am assured that my

government will support me should I require such support, and even now, while the ambassador at Constantinople has been instructed to advise the Porte in favour of the enterprise, Count Walewski informed Fuad Pasha, previous to his departure for London, that the Emperor took particular interest in the Suez Canal, and was anxious to see the Sultan give a token of initiative and independence in the matter, and that the course which Turkey had so far pursued in the matter was, in fact, felt by France to be ground for just complaint."

To M. de Regny, Interim Agent in Egypt.

"Paris, *January* 1, 1859.

"The constitution of the financial company, which will carry out the making of the Suez Canal, has brought the year 1858 to a very satisfactory close, but we must be prepared for a struggle even more severe than any of those which have gone before, for the hostility of the English Government seems to have been exacerbated by the success of our subscription. Our adversaries are beginning to reproach me with having composed the administration exclusively of relatives and friends, to the exclusion of great financiers, but my reply to this is that one gets on best in business with friends and not with enemies, and that to fight these latter I could not well select my colleagues to suit their convenience.

"Then, again, they are trying to undermine the confidence of my supporters by dwelling upon the

risky character of a company which has not got that
wonderful firman which England alone prevents being
issued, and by asserting that the company is irregular in its constitution because Great Britain and
other countries are not among the subscribers.

" My report to the Viceroy, dated December 31st,
1858, has given him a full account of the board meetings held since the constitution of the company was
duly declared. His Highness having wished that the
French investments should not much exceed one-half
of the whole, in order that the company might, so far
as possible, maintain its universal character, we have
fixed the total number of shares as follows :—

Name of Country.	Number of Shares.
France	207,111
Ottoman Empire (inclusive of the Viceroy's personal investment)	96,517
Spain	4,046
Holland	2,615
Tunis	1,714
Piedmont	1,853
Switzerland	460
Belgium	324
Tuscany	176
Naples	97
Rome	54
Prussia	15
Denmark	7
Portugal	5
Sums held in reserve for the subscriptions from Austria, Great Britain, Russia, and the United States, which the Viceroy authorizes me to guarantee for him should they not be taken	85,506
Total number of Shares forming the capital of the Company	400,000

"Thus it is made very clear that I have not attempted to monopolise for France and Egypt the merits of the subscription which, despite all that may be said or done, will not fail to be universal in its results.

"My last news from England is to the effect that we shall get no money from there. The utterances of Lord Palmerston and Stephenson, the engineer, have told. But as we shall go forward, despite the policy of our dear allies, I am not sorry to succeed without their financial assistance, and notwithstanding their hostility, just to take down a little of their insular presumption, accustomed as they are to regard everything impossible which has not their support.

"I forward you the summary remarks of the engineers of the International Commission to the declarations of Mr. Stephenson. With regard to those of Lord Palmerston, totally devoid of reason as they are, his successors will persevere in the same hostile course. I know, through my friends in the foreign corps diplomatique at Paris, that since the success of our subscription the English Cabinet has made redoubled efforts to create difficulties for us with other Powers.

"Thus, for instance, the Marquis de Villamarina, Sardinian Minister in Paris, has been asked by the English ambassador to inform Count Cavour that the English Government was still very opposed to the canal, and that, as matters stood between England

and Piedmont, it would be very detrimental to the future of the latter state if it compromised itself by running counter to English policy.

"I know, too, through Italian friends, that the same intimation was made direct to Count Cavour through the British agent at Turin.

"According to a letter from New York, I must not now count upon any shares being taken in the United States. It will probably be the same in Russia, owing to the financial embarrassment of that country.

"With regard to Austria, the information sent by Bruck and Revoltella, continues to be favourable, despite the death of our good and trusty friend Negrelli. I propose to visit Vienna and Trieste on my way to Egypt next month."

M. de Regny to M. de Lesseps.

"ALEXANDRIA, *January* 2, 1859.

"I send you a brief account of an interview which has just taken place between the English consul in Egypt and the Viceroy. The importance of this interview cannot be exaggerated, for just when an effort is being made to get the world to believe that he is unfavourably disposed towards the enterprise, he replied with remarkable firmness to the English agent that this was his work, and that he was resolved to go on with it, as the Hatti-Sherif of 1841 unquestionably gives him the right to do. We shall see whether the action of the consul is countenanced by his government. It is entirely out of character with the principles of humanity and commerce, of which the English claim to be the principal exponents.

"The facts are as under. Mr. Green went to Cairo on December 11th, and pointed out to the Viceroy that by having granted M. de Lesseps the concession he would find himself exposed to much annoyance, and that M. de Lesseps, upon the strength of this declared that he had your mandate, and had constituted a company. The consul added that no doubt his Highness would repudiate this statement as to your having his mandate.

"Said's reply was : ' People are mistaken in Europe if they attribute the piercing of the isthmus to M. de Lesseps alone, for I am the promoter of it. M. de Lesseps has merely carried out my instructions. You will ask me perhaps what my motive has been, and I will tell you that it has been to bring honour on my name and serve at the same time the interests of the Ottoman Empire. I have acquired by this means the sympathies of all the nations of Europe. You are aware that most of the great Powers are interested in the making of the canal.' The consul replied : 'May I point out to your Highness that if it has been approved of by France and other Powers, it has been strongly opposed by the English Government as contrary to its interests.' The Viceroy said that he was resolved to do all he could to accelerate a work which was generally desired, and gave Mr. Green permission to report their conversation to his government.

" The Viceroy was all the more justified in making this outspoken reply to the irregular step taken by Mr. Green, seeing that he has just completed for the benefit of England the railway from Alexandria to Suez. He deserved some better return for the outlays he has made in English factories and workshops, notably in those of Stephenson, and for the rapidity with which the works were carried out."

To the Duc d'Albuféra, Vice-President of the Suez Canal Company.

"VIENNA, *February* 21, 1859.

"I have already had a long conversation with Baron de Bruck and his colleagues in the Ministry. We are quite agreed as to the subscription for shares being announced in all the towns of the empire, under the patronage of government. Each country has its usages, and it appears that here a public appeal for funds would not answer. I am going to-morrow to Trieste, where deputations are to wait on me, and where I shall arrange with M. de Revoltella for realising the Austrian subscription for shares.

" The venerable Prince Metternich greeted me, as was his wont, with extreme good nature, and complimented me upon my 'manipulation de l'entreprise de Suez,' to use his own words, adding that if we went on steadily and prudently, the irresistible force of truth made our success certain. I am going to jot down our conversation, which was a very interesting one, in my journal."

To the same.

" ALEXANDRIA, *March* 7, 1859.

"I have presented the deputation from our board to the Viceroy, and handed him the declaration, of which I enclose you a copy. After the customary compliments, I had a private audience with the Viceroy, being anxious to see what impression the recent visit of the English consul had produced upon his mind. I found him as kind as ever for me, and thoroughly resolved to pursue, *or perhaps rather to let me pursue*, the enterprise of the canal. He confirmed the accuracy of the report of the interview sent us by M. de Regny, but added that the consul had at the same time thanked him for the completion of the railway, which, to use the expression contained in a letter of congratulation from the P. and O. Company, ' so happily realises to the advantage of England the wished-for communication between the Mediterranean and the Red Sea.' The Viceroy afterwards asked the French Consul-General whether he would, if necessary, support the operations of the canal company? M. Sabatier replied that he had no instructions, but would apply for them if required. The Viceroy's secretary then went to see M. Sabatier, and officially requested him to inform his government of the step which had been taken by his English colleague, and of the embarrassment which he felt in consequence of this persistent pestering. M. Sabatier has asked for instructions by this post, but in the meanwhile has not thought it right to give the Viceroy any advice, for which I do not blame him, considering that I, though in no official position like him, have not thought it right to ask his Highness to intervene, no ostensible act of hostility against the company having yet been committed.

"The Viceroy informed the Austrian consul and myself that *no difficulty had been raised by the Porte,* to which he was about to report what had occurred, but that the opposition came entirely from the English agent. At the same time, I am about to proceed, by arrangement with the Viceroy, to carry out the decisions of the board so far as concerns the continuation of the preparatory survey, works which would in any case have had to be done first of all.

"The other questions are settled in principle, but we must, of course, wait to see what attitude the French consul will be ordered to assume."

To M. Damas-Hinard, Private Secretary to the Empress of the French, Bayonne.

"LA CHÉNAIE, *October* 7, 1859.

" Our ambassador at Rome writes :—

" 'I am following with intense interest the grand enterprise to which you are so patriotically devoting your persevering efforts, and I sincerely trust you will succeed. I know of no more national or useful work than yours.'

"This will give an idea of what our diplomatists think of the Suez Canal. It may be added that the Duc de Gramont, in writing thus, is the mouthpiece of the Roman Court, and the whole of the Catholic clergy is deeply interested in the execution of the work. Only the other day, the Bishop of Orleans, in a pastoral letter, expressed his most ardent wishes for its success.

"The army has, with its usual spirit, taken up our enterprise, and many officers of all ranks are among our shareholders.

"The intervention of the Emperor, which now becomes a question of life and death for us, will certainly increase his popularity at home and his influence abroad. All the governments are ready to support him against the isolated opposition of the antiquated policy of England. This homage rendered to the Emperor's political ascendency reminds one of that paid to Charles VIII. when a battle was about to be fought. The nobility opened their ranks, and, leaving him the foremost place, said : ' To your Majesty be left the honour of making the first thrust with your lance.'

"I said at the last meeting of our board that the Empress had been our guardian angel, and that she would be for the union of the two seas what Isabella, the Catholic, was for the discovery of America. We have therefore chosen the 15th of November, the feast of St. Eugénie, for our first general meeting of shareholders."

To the same.

"LA CHÉNAIE, *October* 13, 1859.

"I learn that the Porte, yielding to the pressure of the English ambassador, has despatched Muktar Bey, the Minister of Finance, to advise the Viceroy as to what course he should pursue in regard to the Suez Canal. If I am rightly informed, his instructions are to discourage rather than stimulate the Viceroy. You will observe that our adversaries, whose motive is easily guessed at, select the time when the general meeting of shareholders has been announced as about to be held, to carry out a threat which will, as they hope, have the effect of shaking the confidence of our friends and create us fresh difficulties. My letters from Alexandria tell me, in fact, that our adversaries, advised beforehand of Muktar Bey's mission, do not make any secret of their belief that *it is all over with the canal, with which the Imperial Government will not have anything to do, leaving the field free to the opposition of the English agents.*

"As her Majesty the Empress will readily see the significance and gravity of these fresh complications, I shall be very much obliged if you will submit this letter to her. She will see how indispensable to me just now is the support she has already so freely given me."

To Mr. D. A. Lange, London.

" LA CHÉNAIE, *October* 15, 1859.

"The *Isthme de Suez* newspaper will give full particulars about the mission of Muktar Bey to Egypt. It is due, beyond all doubt, to the intervention of Sir H. Bulwer, and I have information to that effect, which comes from the fountain-head. The French ambassador at first remonstrated against this mission as hostile to the Suez Canal, but the action of the English ambassador was of such a character that a grave conflict might have ensued; so the French ambassador, in compliance with his general instructions, which are to avoid anything of the kind, left the field free to his English colleague. You may rely upon this information, and the occurrence is a fortunate one for us, as no doubt that was what the Emperor was waiting for, to inform Lord Cowley that he intended to support us, and that the demands of the company must be complied with. In fact, a despatch to this effect has been sent to our ambassador in London, requesting him to communicate it to your government.

"I regard our cause as won, seeing that the Emperor takes it under his protection."

To M. de Ruyssenaers, Alexandria.

"PARIS, *October* 24, 1859.

"I am pleased to inform you that we were received by the Emperor at St. Cloud yesterday. MM. Elie de

Beaumont and Baron C. Dupin, our honorary presidents, joined us, and we were most kindly greeted by the Emperor, who was aware of the object of our visit, and who, speaking to me, said, ' How is it, M. de Lesseps, that so many people are against your enterprise ? ' To which I replied at once, ' Your Majesty, it is because they think you will not stand by us.' The Emperor, twisting the tips of his moustache with his fingers, as he is in the way of doing when he is thinking of what he shall say, observed, after a brief silence, ' Well, do not be uneasy. You may count upon my assistance and protection.'

" Speaking of the resistance of England, and referring to a recent reply of the London Cabinet, which he called a ' startling ' one (*raide*), he added, ' It is a gust of wind. We must take in sail.'

" We then asked him to authorise us to announce to our shareholders that as *negotiations were in progress the general meeting would be adjourned*, as otherwise we should be obliged to refund them their money. He assented to this, and also to our letting it be known in Egypt that he had already given his Minister of Foreign Affairs orders that our rights and operations were to be upheld. We thanked him for it, but we complained of the conduct of the French Consul-General in Egypt, who had entirely failed to protect our interests, and handed a written memorandum in support of our statement. Thinking it time to leave, I made a sign to my colleagues, and finally

observed that I thought it desirable that I should go to
Constantinople and Alexandria, to which the Emperor
replied, 'It is very important that you should do so.'

"My colleagues then retired, but having remarked
that the Emperor wished to speak to us, the Duc
d'Albuféra and myself remained behind. The Emperor
then said to me in a very friendly tone, 'What do
you think we should do now?' I replied, 'Your
Majesty, I think it would be wise to recall the French
Consul-General, who, being a man of great capacity,
could be sent to some other post.' 'Well,' remarked
the Emperor, 'if that is all, it is easily done. You
can tell Walewski so.'

"I lost not a moment in writing to Count Walewski, to tell him what had passed, and I ended my
letter by saying:—

"'The practical result of this audience seems to be
that, while reserving the political question, which can
be left for diplomatic settlement, M. Thouvenel should
be instructed to ask the new Grand Vizier (who is, I
believe, favourable to the enterprise) for a letter to
the Viceroy, authorising him to continue the preparatory works as defined in my letter from Corfu on
the 3rd of March to the Grand Vizier, and, secondly,
that M. Sabatier's services should be utilised anywhere else than in Egypt.'

"It is very fortunate that I happened to be in
France, and not in Egypt, during the mission of
Muktar Bey, which has occasioned you so much

annoyance, and in connection with which you have
given so many proofs of your tact and devotion to the
interests of the company."

To Count Th. de Lesseps, Paris.

" CONSTANTINOPLE, *November* 23, 1859.

" Our minister was well advised in sending me
here, for though at first Thouvenel was rather alarmed,
for fear that some complication should arise in the
midst of his Montenegrin negotiations, I regard my
stay at Constantinople as being most opportune just
now.   I should add that, owing to bad weather, the
letters which ought to have arrived a week before had
only just been delivered when I came, so that Thouvenel had scarcely had time to read them, and feared
that it would be very difficult to bring about a sudden
change of front.   But this also I regard as a fortunate
circumstance, and, moreover, we soon got on capitally.
But he must be well backed up from Paris.   Don't let
them be afraid of the struggle with Sir Henry Bulwer, who, though a personal friend of mine, thinks it
his duty as a good Englishman to serve his government, right or wrong, for which I cannot blame him.
He was confined to his bed with fever when I arrived,
but my presence had the effect of a good dose of
quinine on him, for he was busy at work the next
morning.   His method of proceeding is to show the
Turks letters from London, in which are described
imaginary conversations between Lord Cowley and

Count Walewski in Paris, according to which the latter had promised England not to support the canal scheme, and this *subsequent* to the Muktar Bey mission. There cannot be a word of truth in this, which is a very old dodge. The dragomans of the English Embassy are instructed to alarm the ministers of the Porte by telling them that their assent to the canal may give rise to a war between France and England, which, whatever its result, would be fatal to Turkey. I endeavour to make them see that, on the contrary, if the Porte hesitated to come to a decision there would be far more danger of France and England being brought into conflict.

"There is another point to which I would also fain draw Count Walewski's attention. When the French ambassador here opens the attack, and is seconded, as he will be, by the representatives of Austria, Russia, &c., it is essential that all our forces should be concentrated on the one point we are endeavouring to attain, and that all other questions should be deferred.

"The dragomans of the English Embassy tell the Porte that Lord John Russell's instructions betoken quite as much hostility to the canal as those of preceding foreign secretaries."

To the same.

"CONSTANTINOPLE, *November* 30, 1859.

"I receive a letter from Paris in which I am told: 'Your enemies—and you must not think that you have not plenty of all sorts—have endeavoured to compromise you in high places with reference to your alleged political opinions. There has been a talk of intimacies, relationships, and even affiliations.'

"I confess that accusations of this kind do not trouble me much, but, on the contrary, I am rather pleased to find that those who have an interest in injuring one who has never done an injury to any man, are obliged to have recourse to such weapons of the imagination. For my official career for the last thirty-four years, and my private life, of which an august personage happens to know a good deal, put me beyond the reach of such wretched calumnies.

"My whole life has been spent in the service of my country, nor have I ever meddled in home politics. I have never once set my foot, even out of curiosity, in a public political meeting of any kind. During my thirty years' consecutive employment abroad I was only four times on leave in Paris, and I was not present at the revolutions of 1830 or 1848. Put out of active employment, upon my own demand, in 1849, and receiving no pay or pension, I devoted myself entirely to my family, and succeeded in making good the inroads upon my small fortune caused by the expenses of my latest missions abroad.

"Sustaining in 1854 a very severe domestic affliction,* I set myself to work upon a project which theoretically had engaged my attention for many years.

---

* Note of the Translator.—M. de Lesseps is referring to the death of his first wife.

Since then there has been no secret about a single one of my actions, and there is nothing in my sayings, writings, or doings, to justify an attack which I should not condescend to notice, but that I was afraid of its just now being detrimental to the success of our enterprise. Read this to Count Walewski, and communicate it, if you think fit, to M. Damas Hinard, for the Empress. She knows that though I did not vote for the empire, I am no factionist, and that though I am a lover of liberty, I am not one of those who would seek to overthrow the order of things which my country has raised up."

To M. Ruyssenaers, Alexandria.

"CONSTANTINOPLE, *December* 7, 1859.

" After several ministerial councils, which resulted in considerable discussion owing to the innumerable steps taken by Sir Henry Bulwer, the Porte agreed to the demand made by the French ambassador. That is to say, a reference will be made to the Powers to cover the political responsibility of Turkey in regard to the canal, and to settle the international questions arising out of it. All that now remains is to decide in what form the reference shall be made. I of course leave M. Thouvenel to take action in his own way, and have not made any move personally. Sir H. Bulwer sees so clearly that this appeal would put an end to all possibility of further resistance that he is moving heaven and earth to prevent it being made.

"If all terminates as I hope, I shall hand to the heads of each Legation a memorandum which I have prepared with confirmatory documents appended."

To Chevalier Revoltella, Trieste.

"CONSTANTINOPLE, *December* 10, 1859.

"There have been two councils within the last three days, and there will be another to-morrow. The Porte is still hesitating, for Sir H. Bulwer has held out threats of war, but we have made the Turkish ministers understand that this is only bluster, and that he would look very foolish if asked to put this in writing."

To M. Ruyssenaers, Alexandria.

"CONSTANTINOPLE, *December* 24, 1859.

"M. Thouvenel has informed me that at last the agreed reference to the Powers has been drawn up, after sixteen ministerial councils. This reference, the terms of which were so long discussed, has been communicated to M. Thouvenel, and by him sent on to Paris.

"The Sultan sent, the day before yesterday, for the Grand Vizier, as well as for Fuad Pasha, the Minister of Foreign Affairs, to congratulate them upon their conduct of the negotiations.

"Yesterday afternoon, we were thunderstruck to hear that the Grand Vizier, Kuprisly Pasha, had been dismissed, and I was afraid at first that there had been a change of front in Turkish policy. But I was at once reassured upon hearing of the nomination of Ruchdi Pasha, whom I lost no time in going to see, and who appeared most friendly.

To the Duc d'Albuféra, Paris.

"CONSTANTINOPLE, *December* 28, 1859.

"The change of Grand Viziers has not in any way affected the situation as I described it to you in my previous letters. The Sultan gave his full approval to what had been done by the ex-Grand Vizier, so I leave to-morrow for Alexandria, where I shall not remain long, as all I want to do there is to see the Viceroy.

"M. Thouvenel is anxious that I should get to Paris as quickly as possible."

To the same.

"ON THE NILE, BETWEEN MONFALOUT AND SIOUT,
"*January* 6, 1860.

"The Viceroy was waiting for me at Monfalout previous to going up the river to Siout. We had a very interesting conversation, and I can assure you that we are perfectly agreed. He is very anxious that we should, without making any fuss about it, at once proceed to the setting up of our dredging apparatus, to the excavating of our service trench (*rigole de service*) as far as Lake Timsah, and to the preliminary works in the inner fort.

"He is very satisfied with the result arrived at at

Constantinople, without his rights having been infringed upon or called in question, and he admits that his rights are our rights.

"I explained to him how his running account stands, and left him a copy of it to examine.

"The Viceroy assured me in the most gracious manner that at no time had his confidence in me ever been the least shaken, and that he was sure he could say the same of me. He repeated what he had already said at our last interview, that we can understand each other thoroughly even when parted. He is very pleased that the French consul has been changed. After our conversation we went up to Siout, each on our separate steamer, and he told me that he would not hear of my leaving."

To M. Ruyssenaers, Alexandria.

"ALEXANDRIA, *January* 11, 1860.

"In handing the Viceroy his account with the company, which he found correct, I pointed out to him that his Treasury had not included in its advances several large sums which his Highness declined to receive, and I thanked him on behalf of the shareholders. These sums related to the surveys made several years ago, the salaries of all the engineers placed at our disposal, the cost of the International Commission during its journey through the isthmus, and many other items."

To His Highness the Viceroy of Egypt.

"PARIS, *January* 26, 1860.

"I arrived here four days ago, and I hasten to send to your Highness, as promised, a copy of the communication made to the French Minister of Foreign Affairs by the Turkish Ambassador. There is no need for me to tell your Highness that this note, which is symptomatic of how the Porte sways to and fro between France and England, does not effect any precise settlement. It is no more than a mere official subterfuge, and it, in short, leaves to time and to the course of events to bring about a definite arrangement which the Porte has not ventured to make. This is a political burial of the question which enables us to act and to force on the solution afterwards. This is what the Spaniards call *cubrir el espediente* (saving the appearances).

"The Emperor has received M. Béclard, the successor of M. Sabatier, and has specially commended to him the interests of the company.

"In agreement with M. Thouvenel, I have obtained from the committee the vote of the resolution of which I enclose a copy, so that your Highness may not be in any way troubled by inquiries with regard to the works we are executing for the creation of Port Said and of the inland fort at Timsah.

"I have seen King Jerôme and his son Prince Napoleon and the Ministers, but I have waited until fully informed on all points before asking for an audience of the Emperor, which I shall do to-day or to-morrow. I send your Highness the model of an apparatus for letting the captains of ships know when the lighthouse at Port Said is lighted. This light-house will be very useful for vessels plying between the coasts of Egypt and Syria."

To the same.

"PARIS, *May* 16, 1860.

" I have the honour to inform your Highness that I shall leave Marseilles on the 18th to lay before you the resolutions passed at the general meeting of May 15th, and to point out to you the satisfactory results which this meeting will have upon the realisa-tion of our enterprise."

To His Excellency Kœnig Bey, Secretary to the Viceroy.

"ALEXANDRIA, *June* 27, 1860.

" I send you a letter from Constantinople, which please read to the Viceroy, whom I shall not see to-day. The best answer we can make to our adver-saries is the arrangement we have concluded with Ragheb Pasha, which, far from being a cause of financial embarrassment, will strengthen the Vice-roy's credit.

" This letter, dated June 10th, is as follows:—

" ' The *mot d'ordre* of the English Embassy, in

public and in society, with reference to the Suez
Canal, is this : " As M. de Lesseps and his shareholders are indifferent to the ruinous impossibility of
the work, which *The Times* has pointed out, so much
the worse for them. It is not England's business to
preserve them from the consequences of their own
folly. It would be absurd to oppose the execution of
a thing which is not possible; and if, by dint of
money expenditure and by ruining two or three
generations of shareholders the canal is made, so
much the better for England—which will derive more
benefit from it than anyone else—and for the integrity of the Ottoman Empire, from which Egypt,
rendered inviolable by the universal interests attached
to the canal itself, will be in no risk of being
separated." '

" Is this a more or less honourable mode of beating
a retreat, or is it not rather, as I believe, an expedient for putting the French Embassy to sleep, and
for making a redoubled attack against the Viceroy?
It is represented that he has wasted and ruined the
finances of Egypt, and that it is desirable to replace
him. Not being able to attack the canal outright, an
endeavour is being made to discredit the financial
position of the Viceroy with respect to the work,
which will, however, cost him much less than the
railway to Suez. Be this as it may, I know that the
Embassy is upon its guard, and that despite the good
will for Sir Henry Bulwer with which Turks said to

be in the confidence of the Viceroy are credited, nothing will be done either against his Highness or yourself. It is always well, however, to keep one's weather eye open.

" M. de Lavalette seems very easy in his mind, and I am bound to believe that he has his reasons for this. One thing certain is that he is determined, if necessary, to display great zeal in an affair in which so many legitimate interests are involved."

To Count Th. de Lesseps, Paris.

"ALEXANDRIA, *December* 28, 1860.

" It may interest the Minister of Foreign Affairs to hear something about our relations with Abyssinia, a country which, now so far off, will, when the Suez Canal is open, be the nearest to Europe of all those on the east coast of Africa. Our consul at Massowah transmitted me a letter from the King of Abyssinia, which has already been published;* but I enclose you the translation of a second letter from him, in reply to what I wrote in answer to his first letter :—

Second Letter from King Nikas Negoussié to M. Ferd. de Lesseps.

" ' Peace be with you !

" ' Your letter duly reached me, and I thank you

* Note of the Translator.—This letter is included in the chapter on "Abyssinia." See Chap. XI.

for the good wishes you express towards me and for the prosperity of my people. I am convinced that, despite my earnest efforts to remove from my country the barbarous customs introduced into Abyssinia during the last few centuries, I shall never succeed in entirely changing the ideas of the people, and regenerating them until European genius, uniting the waters of the Red Sea and the Mediterranean, has opened our country to European commerce and Christian civilisation.

"'When, by the grace and will of God, I have brought all the rebels into subjection, and established my kingdom upon a more solid basis, I shall send my ambassadors to all the Christian kings of Europe, and I shall say to them: "My brothers, I am like you a servitor and a son of Jesus Christ. Receive me, therefore, among you and enter into relations with me. Then the men of Europe will come among us, they will teach us your arts, and Abyssinia will become what it was before. If you are my friends, and if you desire the good of humanity, you will doubtless aid me in this work."

"'In order that my acts should correspond to my words, I have forbidden the mutilation of the killed or wounded in battle; I have prohibited the slave trade, and this odious traffic is now suppressed in the Tigré, Semen, and all the subject provinces upon the coast of the Red Sea.

"'Permit me to repeat to you in conclusion that

you can count upon me for anything which is calculated to advance the work of the canal.

"'May the good Lord keep you.'"

"'DENEAT AXOUM, 8 *Lasoli*, 1852.'

"I beg you also to hand to Count Walewski,
together with my correspondence with King Nikas,
my *Abrégé de l'Histoire d'Abyssinie* as likely to be
useful for purposes of reference in the political relations which will certainly follow the opening of the
Suez Canal to navigation between Europe and the
east coast of Africa."*

My Journal.

"*17th, 18th, 19th, 20th January*, 1863.

"Having left Ismailia on horseback in order to
reach Kantara more quickly than I could by water, I
cross the desert, followed by my faithful Hassan, the
night being dark and there being nothing but the
north star to guide us. After two hours' repose, I
am awoke by a courier, and on opening the despatch
I find that Mohammed Said, who was very ill when
he reached Alexandria, is in a very critical state, and
that if I wish to see him again there is not a
moment to be lost. I have a horse saddled, and,
instead of taking the desert route, I determined to

* King Nikas's intentions were not carried into effect, for he was
treacherously betrayed to the Pretender Theodoros, who had him
cruelly put to death, and governed Abyssinia until his barbarity
to foreigners led to the English expedition, under Lord Napier, the
capture of Magdala, and his suicide.

follow the banks of the canal and gain time. There are several solutions of continuity, but my horse gets me out of all the difficulties, and I arrive at Ismailia at break of day. I had telegraphed in advance to have a bark got ready, with two dromedaries to draw it along the banks, but just as I reached Tel-el-Kebir, I meet another bark which was bringing up Jules Voisin, who had been sent by M. Guichard, director of our domains at Ouady, to tell me that the Viceroy had died on the morning of the 18th. I am grieved to the heart, not on account of my enterprise, in which I have the most serene confidence, despite all the difficulties which may arise, but because of the cruel separation from a faithful friend who for more than a quarter of a century had given me so many proofs of affection and confidence. As I travel on to Alexandria, I go over in my mind all the circumstances connected with our youthful friendship, his careless and easy life as a young man, and his beneficent reign. Before seeking a little repose I ask permission from the noble and estimable princess, his widow, to allow me to enter the family mosque in which his body had just been lodged. I remain there an hour quite alone, with my head resting upon the dead man's turban. His servitors, whom I afterwards question, inform me that towards the close of his illness their master used a stick which I had given him while we were on an excursion together, and that he had it at his side when he died. I have every search made to discover this relic, but it is nowhere to be found. I have a description of it given to the police, who eventually discover it in the hands of an Arab as he walked along the street. It was restored to me, and the history of this interesting souvenir is as follows :—One day Mohammed Pasha, upon my return from England, showed me two sticks, the one which I had given him and one which was a present from an English admiral, and said : ' You sometimes mention the canal business to me in the presence of persons who might repeat our conversation at an inconvenient moment. To obviate this, whenever you come to see me and you notice that I have the English stick, you will remember that nothing is to be said about the canal; but you can say as much as you like when you see that I have your stick.'

"After remaining three days at Alexandria, and giving time for the official congratulations offered to Mohammed's successor to be got over, I start for Cairo, where the new Viceroy, far from being offended, expressed himself much pleased at the regret which I expressed and felt, and of his own accord assured me that he would treat the widow, son, and household of his predecessor as if they belonged to his own family."

To the Duc d'Albuféra, Paris.

"Cairo, *January* 24, 1863.

"Summoned by telegraph when the Viceroy was dying, I reached Alexandria from Kantara in twenty hours, but too late to close the eyes of one who had ever been for me a firm and fast friend. The new Viceroy, Ismail Pasha, has been pleased to give me his assurance of goodwill towards our enterprise, as I telegraphed to you ; and I am now, after having had a long and confidential conversation with him, in a position to assure you that we may feel quite at ease both as regards the progress of our works and the regular payment of the sums for which the Egyptian Government has made itself responsible. Ismail Pasha is opposed to the idea of a loan, if it can possibly be avoided, and he is anxious, if possible, to have all the instalments paid in succession, so as to enable the company to meet all its expenses without having any need to make a further call upon its shareholders until the whole debt of the Egyptian Treasury has been paid off. We intend to draw up a plain agreement to this effect on the Viceroy's return from Constantinople, where he is about to go to receive his investiture from the Sultan. Until then it is easy to understand that Ismail Pasha cannot do more than let things remain in the state in which they were left by his predecessor, but I am assured by him and his intimate friends that he understands how important it is for the glory of his reign to bring the enterprise of the Suez Canal to a successful conclusion.

"The Duc de Brabant, who has returned from an excursion in Upper Egypt, has expressed to me his wish to visit our works in detail, and I am starting with him this morning, the Viceroy having ordered a special train for us from Cairo to Samanoud and a steamer from Samanoud to Damietta. I have telegraphed to M. Voisin to meet us, for before I knew of the Duc de Brabant's proposed visit we had arranged to inspect our works together.

"The Viceroy will return from Constantinople in about three weeks, and we shall then make our financial arrangements previous to my starting for France, and he has repeated to me several times, ' I don't wish you to reach Paris until the company is completely satisfied.' He made a similar declaration to our consul, and also told him that he intended to effect the payment of his shares in such a way as to obviate any necessity for making a fresh call upon the French shareholders.

"His Highness informed me a few days ago that he had steamers to bring contingents of workmen from Upper and Middle Egypt for the month of Ramaden, during which period there is not, for this once, to be any suspension of labour. It was very desirable that such should be the case, as an interruption of the work would certainly have been misinterpreted, and this the Viceroy saw.

"These facts confirm, therefore, the favourable dispositions which his Highness manifested from the first, and our affairs in Egypt are going on as well as possible."

To the same.

"ALEXANDRIA, *March* 10, 1863.

"I took care to be at Alexandria upon the return of the Viceroy from Constantinople, and I was one of the first to see him. He told me in confidence all that had occurred during his visit to the Sultan, as you will learn from my brother Theodore, whom I have requested to communicate them to you before inform-ing M. Drouyn de Lhuys of them.

"The Viceroy's voyage has produced the best pos-sible results for us, and, to use his own words, he said to me, 'If you had been Viceroy of Egypt as well as president of your company, you could not have done better in the interests of the Suez Canal scheme.'

"There need, therefore, be no fear now as to the rapid progress of our works, and the discharge of the debt due from the Egyptian Treasury. The Viceroy started yesterday for Cairo, after receiving the new French Consul-General, my old friend M. Tastu, who will do all he can for us, though we must not forget the services rendered us by M. de Beauval."

To Count Th. de Lesseps, Paris.

" CAIRO, *August* 28, 1863.

"I have just received from an intimate and devoted friend in Paris the following letter :—

" 'I think it right to let you know what I have

just heard, and you will be the best judge as to what it is worth. I can see no harm in letting you know this, for if there was the slightest foundation for it, it would be very unfortunate if you were not fore-warned. The information was given to me on the express condition that I should not disclose to you the source from which it came. It appears that a head engineer of the Ponts-et-Chausées was sent to Egypt by a statesman now in power, with the mission to inspect the works on the isthmus, and address him a report upon the results of his inspection. I am told that this person expressed himself very unfavourably as to what he had seen in the course of his visit, and that he was very severe upon your engineers. It is considered certain that his report will be very hostile, and that he will draw the conclusion that the affair cannot possibly be carried through under present con-ditions. It is anticipated that this report will be handed to the statesman in question, and that he will submit it direct to the Emperor. Armed with this report, the person who presents it will endeavour to persuade the Emperor that the affair is being badly managed, that the capital of the shareholders is in danger, and that the honour and success of the enter-prise is at stake; while, by way of fresh arguments to use with the Empress, from whom more difficulty is anticipated than from the Emperor, an effort will be made to alarm her and to persuade her that, in *your interest*, it is desirable to save you from the difficulties which you are heaping up for yourself. The object is to bring about the liquidation of the present company, and substitute for it another which is already in course of formation. There is some talk, in addition, of another company composed of large bankers.'

"If I were in Paris my first step would be to show the statesman in question the letter I had received. I should ask him to request the engineer if he had made any observations more or less favourable to the course of our works, which had been directed by his colleagues of the Ponts-et-Chaussées, to communicate these observations, so that we might have them controlled and verified by four of the most distinguished of his colleagues, MM. Tostain and Renaud, inspectors-general, and the engineers MM. Pascal and de Fourcy, who are just coming out to Egypt.

"With regard to the inheritance of the Suez Canal, it is not upon the point of being divided; we have given sufficient proofs of being alive, and we are, thank God, in pretty good health. Our first steps were attended with difficulties, and our childhood was a stormy one, but we have reached the age of manhood. We intend to prove that, if we have been able to constitute ourselves financially, without the assistance of great capitalists, so, with the help of able engineers, we shall be able to complete our work, without delivering ourselves to great speculators, who would not be sorry to absorb a part of our share-

holders' money.   We have laboured and sown; we
intend to reap the harvest.

"These fresh intrigues, if they really exist, will
share the fate of the financial and political intrigues
which have preceded them.

"I tell you what I think, and must leave you to
decide as to whether it is expedient to inform the
Empress of the matter."

*To His Highness Prince Ismail, Viceroy of Egypt
and Ethiopia.*

"CAIRO, *September* 1, 1863.

"Monseigneur,—A letter from the Grand Vizier
was addressed to your Highness in the early part of
August with reference to the Suez Canal.

"The French Embassy at Constantinople having
succeeded in obtaining a copy of this letter, and communicated it to me, I have lost no time in drawing
out a memorandum on the subject, in which I venture
to call your close attention.   I may at once say that I
am of the same opinion as the French Government,
which has never, it is true, had occasion to take any
initiative in regard to the Suez Canal, and which has
rightly refused to make a political question of it, but
which is firmly resolved to uphold, together with
your rights, those of the company in which French
capital has been legitimately invested.

"It will be for the representative of the Emperor
at your Highness's Court to give you, with more

authority than myself, the same assurances, and to encourage him, upon the other hand, not to permit any interference in the internal administration of Egypt contrary to the arrangement of 1841, which constituted the Egyptian Power in favour of the line of Mehemet Ali.

" I trust that your Highness, whose protection and aid have been so freely accorded me since the beginning of your reign, and who is more interested than anyone else in the success of the enterprise at the head of which I have the honour to be, will appreciate the obligation which is incumbent upon me to scrupulously discharge all my duties, and that you will help me to employ the necessary means for completing as promptly as possible the work from which you will derive so much glory and profit."

Such is the origin of the work of the Suez Canal.

With regard to the celebrated firman which provoked so many international negotiations, the company went on its way without concerning themselves any more about it, and without a day's delay.

The tranquillity of the president was to a great extent due, especially during the last few years, to a fact which has remained unknown to the public.

When Napoleon III. arrived at Marseilles, on April 30th, 1865, to embark on his yacht, the *Aigle*, on his way to Algeria, the Grand Vizier, Fuad Pasha, who had come to the south of France to recruit his health, was among the crowd of notables who were grouped around the Emperor, who took no notice of him, and did not reply to his bow. He then came up closer and asked the Emperor if his Majesty had any cause of complaint against him or his government. The only answer he got was an expressive gesture accompanying the single word "the firman."

This firman was in the end granted. The grand inauguration of the canal took place on November 17th, 1869, in presence of the Empress Eugenie, the Emperor of Austria, the Prince Imperial of Germany, the Prince of Orange, General Ignatieff, representing the Emperor of Russia, and the ambassadors of all the Powers from Constantinople. The number of vessels which went through the canal from Port Saïd to Suez was sixty, and the multitude of guests—men of science, men of letters, and artists, from all countries—were treated by the Khedive Ismaïl with a magnificent hospitality unexampled in history.

This is a homage which I am proud to pay him after the painful occurrences which have afflicted Egypt and removed him from power.

CHAPTER V.

IT will, I think, not be out of place if I supplement this chapter with "a question of the day" (*actualité*), in the shape of a letter which I addressed to Lord Stratford de Redcliffe in 1855, with reference to an eventual seizure of Egypt, either by France or by England.

"CONSTANTINOPLE, *February* 28, 1855.

"There are questions which it is necessary to face openly, in order to solve them aright, just as there are wounds that must be probed before they can be healed. The straightforward way in which you met my preliminary observations with reference to an affair, to the gravity of which I am fully alive, emboldens me to submit to your consideration one point which, as it seems to me, it is desirable to keep in view with reference to the Isthmus of Suez. Owing to the great influence which your character and your long experience enable you rightly to exercise in the decisions of your government in all Eastern questions, I am specially anxious to omit nothing which

M 2

may assist you in forming your opinion in full knowledge of all the facts.

"The results already obtained by the ultimate alliance of France and England show very clearly how
advantageous their union is in the interests of the
equilibrium of Europe and of civilisation. It concerns, therefore, the future and the happiness of all
the nations of the universe to maintain intact, and to
preserve from any shock, a state of things which, to
the lasting honour of the governments which have
brought it about, can alone, with the aid of time,
ensure to humanity the blessings of progress and of
peace. Hence follows the necessity of getting rid,
without delay, of any possible cause of rupture or
even of coolness between the two peoples. Hence, in
consequence, it was our bounden duty, with a view
to future contingencies, to search out what are the
circumstances calculated to awaken the secular feelings of antagonism, and to provoke, either upon the
one side or the other, any of those emotions against
the force of which the wisest of governments is
powerless to contend. The motives of hostile rivalry
show a tendency gradually to give way to that
generous emulation which engenders great achievements.

"To look at the situation from a general point of
view, one fails to see upon what ground, and *à propos*
of what, the struggles which have so long caused the
world to reek with blood, are likely to be renewed.

Are the two peoples divided by financial and commercial interests? Why, the capital of Great Britain, invested in all manner of French enterprises, and the immense development assumed by international commerce, establish between them ties which grow closer every day. Are political interests or questions of principle at stake? Why, the two nations have but one and the same aim, but one and the same ambition —the triumph of right over might, of civilisation over barbarism. Is there any petty jealousy with regard to territorial extension? Why, they both recognise now the fact that the globe is large enough to offer to the spirit of enterprise which animates their respective populations land to be cultivated and human beings to be redeemed from barbarism; and, moreover, so long as their flags float side by side, the conquests of the one benefit the activity of the other.

"At first sight, therefore, one can see nothing in the general aspect of affairs which can affect our friendly relations with England. Nevertheless, looking at the matter a little more closely, there is one eventuality which, seeing how the most moderate and enlightened cabinets are impelled to share popular passions and prejudices, is capable of reviving ancient antipathies, and of compromising the alliance and the benefits deriving from it.

"For there is one point of the globe, upon the free right of way through which depends the political and commercial power of England, a point which France,

for her part, in centuries past, had the ambition to
possess. This point is Egypt, the direct route to
India—Egypt, which has been more than once dyed
with French blood.

"It is superfluous to go into the motives which
could not allow England to see Egypt fall into the
hands of a rival nation without offering the most
desperate resistance ; but a fact which must be also
taken into full account is that France in her turn,
though not so materially interested, could not, in
obedience to her glorious traditions, and under the
impulse of other sentiments more instinctive than
logical—and for that very reason all powerful upon
her impressionable inhabitants—allow England to
take peaceable possession of Egypt. It is evident
that as long as the route to India is open and
safe, that the state of the country guarantees facility
and promptitude of communication, England will not
voluntarily create for herself the gravest difficulties
in order to appropriate to herself a territory which,
in her eyes, is only valuable as a transit route. It
is equally clear that France, whose policy for the last
fifty years has consisted in contributing to the prosperity of Egypt, as well by her counsels as by the
assistance of a great many Frenchmen distinguished
in science, in administration, and in all the arts of
war and peace, will not, for her part, attempt to
realise the projects of another age so long as England
does not set foot there.

"But should one of those crises which have so often shaken the East occur, or any circumstance arise which should compel England to take up a position in Egypt, in order to prevent any other Power forestalling her, it is certain that the alliance would not survive the complications which such an event would bring about. And why should England consider herself forced to make herself mistress of Egypt, even at the risk of breaking up her alliance with France? For the simple reason that Egypt is England's shortest and most direct route to her Eastern possessions, that this route must be constantly open to her, and that upon this vital point she can admit of no compromise. Thus, by reason of the very position which in nature she occupies, Egypt may again be the subject of a conflict between France and Great Britain, so that this chance of a rupture would disappear if by some providential event the geographical conditions of the Old World were altered, and the route to India, instead of traversing the heart of Egypt, was put back to its limits, and, being open to all the world, could no longer be the privilege of any one nation in particular.

"Well, this event, which must be in the designs of Providence, is now within the possibility of human accomplishment. It may be achieved by human enterprise, and may be realised by piercing the Isthmus of Suez—an undertaking to which nature offers no obstacle, and to which the capital of

England, as well as of other countries, would certainly contribute.

" Let the isthmus only be pierced, let the waters of the Mediterranean mingle with those of the Indian Ocean, let the railway be continued and completed, and Egypt, acquiring greater value as a country of production, of internal trade, and of general transit, will lose its perilous importance as an uncertain or contested route of communication. The possession of its territory, no longer being of any interest to England, will cease to be a possible cause of contention between her and France, the union of the two countries will become henceforward unalterable, and the world be saved from the calamities which would attend a rupture between them. This result offers such great guarantees for the future that the mere indication of it will suffice to command the sympathy and the goodwill of the statesmen whose efforts are bent upon placing the Anglo-French alliance upon immovable foundations. You are one of these men, my lord, and you have such a predominant part in the discussion of great questions of state that I am most anxious to acquaint you with my views and aspirations."

CHAPTER VI.

AFTER THE WAR OF 1870–1871.

IN the year which followed the conclusion of peace
with Germany, the public administrations had
to undertake multifold and contradictory duties, which
created great complications, and entailed expenses
which it is difficult to measure until one comes to
examine them in detail. It was necessary both to
disorganise the war services, to reorganise the peace
services, and to make good the disasters which had
broken up all the machinery of ordinary government.
The first obstacles in the way of a return to a normal
state of things having been cleared away, an immense
amount of labour remained to be done in order to con-
solidate the work of peace.

Public and private interests had been so profoundly
troubled by the ten months of war and internal disturbance, so many transformations were rendered
necessary by the new order of things, the re-establishment of the country was so ardently desired, that an
immense number of laws, decrees, and administrative
measures were passed day after day, so to speak.

There would be a real interest and a patriotic duty in making a compilation of all the acts which were accomplished with the common object of raising the prestige of France, of getting together the scattered documents upon which it would be easy to lay hands to-day, but which will be forgotten to-morrow.

A work of this kind would be not merely the diplomatic history of the peace with Germany, but the history of the reconstruction of our country.   When fate involves a nation in disaster, such as the war of 1870 was, there are two phases through which it passes before resuming its rank in the world : the diplomatic phase of the treaties which regulate peace and its direct and immediate effects; and the longer phase during which the wounds of the war are closing, order is being restored in the country, the truncated limbs of the amputated territory are being tended, the administration and finances are being reorganised, and, in a word, the political equilibrium of the country is being restored.

History has related the main outlines of the events of 1870, and has also revealed certain anecdotal and dramatic details of special interest.   The publications which have hitherto appeared have done little more than register diplomatic documents, and a few official letters, &c., so that I may say a few words about the results of the conventions of 1871.

The diplomatic work done in 1815 was so great and so complicated that it has of itself absorbed the atten-

tion of public writers, for the re-arrangement of territory which took place at that period extended to the
greater part of Europe, and something like a fresh
equilibrium of the Western world came into existence.
In 1870 we had to treat with Germany alone, the rest
of Europe being content to look on. The diplomatic
agreements were, no doubt, less numerous than in
1815, but the political reconstitution of France, which
was recovering, not only from a foreign war, but from
an internal revolution and a formidable insurrection—
one, it may be said, without precedent in her history,
plus quam civilia bella!—necessitated an immense
number of operations connected more or less directly
to peace. As a case in point, let me instance the
making good of the damages arising from the invasion.
Of course, it was impossible to indemnify everyone,
and most of those who received pecuniary grants did
not recover all that they had lost. The whole of the
public fortune would not have sufficed for that, and,
moreover, there are losses which no money can make
good. But the sacrifices which France has made since
1871 for the victims of the war is the best proof of
the progress of civilisation and of national harmony
which have been exhibited since the beginning of the
century. In previous wars, and after those of the
First Empire, it never occurred to anyone that the
citizens of a country, being inter-dependent the one
upon the other, were in duty bound to form a sort of
mutual assistance fund for those who had suffered the

most. The victor alone turned his triumph to account,
making the vanquished compensate his subjects for
what they had lost. It was thus that in 1870, as in
1815, France was crushed by the weight of the
ransoms which she had to pay, but the difference
between the two epochs is that in 1870, despite the enormous liabilities which defeat had entailed, the country
did not forget the provinces which had felt the full
weight of the invasion, and repaired, to the best of its
ability, the damage which had been done there. The
State showed itself liberal in its dealings with foreigners
as well as Frenchmen, both alike being allowed to
profit by the laws relating to indemnities. This
example will not, it is to be hoped, be forgotten by
any foreign countries which may be subjected to a
like trial, and in which Frenchmen may be residing
and may have suffered loss, either from foreign war or
internal discord. For, it must be remembered, indemnities were granted as well for the losses occasioned by the German war as for those due to the
Communist insurrection. These indemnities were not
confined to individual losses, but were extended to
collective and corporate bodies. So it was that large
grants were made to railways; that departments and
parishes were reimbursed for their expenses in connection with the mobilisation of the National Guard;
and that the road bridges destroyed during the war
were rebuilt at the cost of the State. The total
amount spent in this way exceeded £34,000,000.

The two hundred millions paid by France to Germany were in part applied to indemnify the Germans for their losses. From the statements in the German budget, it appears that a sum of £58,200,000 was paid for losses incurred by the war, while a further sum of £58,376,500 was granted to German ship-builders, which may be taken as representing the losses which our navy inflicted upon the maritime trade of the enemy.

The indemnity allowed for bombardment in Lower Alsace amounted to about two and a-half millions, nearly the whole of which was paid in Strasburg. The further employment of the war indemnity which we paid reveals some interesting details. Thus we find that the imperial fortresses received £10,800,000— those of Alsace £6,450,000. The Invalides received £28,033,800, while an imperial treasure of £6,000,000 was created, and nearly half-a-million sterling was spent in rewarding distinguished services. The pensions for soldiers invalided during the war exceeded two millions sterling, while the total losses which the Germans incurred during the campaign amounted to 129,250 in killed, wounded, and missing, of whom 5,153 were officers, 11,095 non-commissioned officers, 1,292 musicians and trumpeters, 595 volunteers, and the remainder private soldiers. There were 44,996 killed; the losses during the first part of the war (July to September) being 74,786, and in the second part (September, 1870, to May, 1871) 54,484.

The battle in which the Germans lost the most men was Gravelotte, where 4,500 were killed and 16,175 wounded or missing.

Reverting to the mode in which the two hundred millions were spent, we find that after deducting the various sums laid out as above, the amount remaining for division between the various German States was £118,411,550, of which the North German Confederation received £79,114,200, Bavaria £13,468,800, Wurtemburg £4,248,200, Baden £3,050,000, and Southern Hesse, £1,400,000.

The payment of the war indemnity to Germany constitutes, with the loans which it entailed, the largest financial operation ever carried out. It was part and parcel of the evacuation of the territory, which was conducted concurrently with it. To form an idea of the manifold constructions and contrivances to which the Treasury had to resort in order to effect the payment of the indemnity, one must read the report of the Budget Committee of 1875, which M. Léon Say presented to the National Assembly. The Bank of France rendered invaluable services in this arduous juncture, but the most remarkable feature of the operation was the international character which it assumed, this being quite a novelty in the economical history of Europe.

All the efforts of all the banking-houses in Europe were concentrated upon this one object. All other business was suspended in order to facilitate the com-

pletion of the French loans and the transmission of
the sum abroad. The French Government did not
pay to Germany in cash more than £21,840,000 in
gold and £10,920,000 in silver, the rest being in
letters of credit and bills. The cost of conversion was
rather more than £500,000, and the only point which
has not been cleared up, and which it would be
interesting to ascertain, is how, after having despatched from France the sums of money collected in
so many other countries, they were then remitted to
Germany, which could only have been done by converting all the other foreign securities into German
securities. It appears that this operation was in a
great measure facilitated by the fact that during the
years 1871-73 Germany was largely indebted to
England for the balance of trade. But the report of
the National Assembly does not give any further details
upon this point.

Another large operation, resulting from the payment of this indemnity, was that which involved the
reconstitution of our war material, and this forms a
chaos into which it is no easy matter to throw any
light, the schemes of the Government and of the
financial committees of the Assembly having varied a
good deal owing to the uncertainty as to what was
the best way to go to work. It is certain that at the
termination of the war, when it was necessary to re-
plenish our emptied arsenals and stores, to reconstitute
our new frontier and our army, there was no means

of including these expenses in the ordinary budget. In 1873 it was decided that the maximum of the expenses to be included under this special heading should be £30,920,000, but this was soon exceeded, and the account was divided into two parts. The first was paid off in 1875, at £36,587,000, while the second, comprising the years 1876-79, absorbed more than £56,000,000. It was only in 1879 that this special estimate could be incorporated in the budget, where it forms an item by itself called, "Dépenses sur ressources extraordinaires." This estimate has necessitated an enormous number of documents, reports, and discussions, which make it very difficult to understand.

One need have a special gift for financial business to make head or tail of it, and M. Villefort's book on the subject may be consulted with advantage, particularly in regard to the accounts of the territory ceded to Germany. At first sight it may appear as if the cession of territory, after a war of conquest, is a matter of public concern only, but we must not forget how many private interests are affected by it and have to be indemnified.

The Franco-German Commission at Strasburg took eight years to effect this settlement, and from their accounts it appears that France paid to Germany for the debts peculiar to Alsace-Lorraine £1,680,000, and received from Germany only £600,000.

The annexation entailed other arrangements, such as

the remodelling of the French frontier departments from the judicial and administrative point of view, and this is not the least interesting part of the whole story. But the main fact, which sums up all the rest, is the total account of what the war cost us. The figures, which tell us this themselves, testify to the financial power and vitality of our country.

The total of this cost, excluding, of course, the losses sustained by the various branches of industry and trade during and immediately after the war, exceeds £1,460,000,000. In this total, extraordinary war expenses are put at about £80,000,000, war indemnities at £36,000,000, and the maintenance of the German troops at £14,000,000. The cost of the different loans is estimated at £25,240,000, and the net loss from the territory annexed at £2,640,000, while the reconstruction of our war and naval material is given at £80,000,000.

The question as to whether the State is responsible to the inhabitants of the country for the damage caused by war is a very important and complex one. Theoretically, it excites the liveliest controversy, and from a practical point of view it forms the subject of constant demands upon the Government. Various views were expressed in the National Assembly, but the majority did not make any exceptions or distinctions which in strict justice could be repudiated. As I have already said, foreigners as well as Frenchmen were allowed to benefit by the beneficent measures adopted, and these measures applied alike to the damage done by the French or the German forces.

The new French frontier has, owing to the division of territory, made necessary a reorganisation of the military and religious services, and here again the various interests which had to be conciliated were most complicated. One of the most difficult matters was the reconstitution of the documents bearing on the identity of the soldiers who had disappeared, and the regulating of their successions, while arrangements had to be made for keeping in order the burial-places of the two armies. The two governments, with much good feeling, agreed that these burial-places should, without distinction of nationality, be kept in a proper state; and at the present time the various spots where the dust of 87,000 Frenchmen and Germans lies mingled together are marked by a funereal monument.

The dead who sleep upon foreign soil should ever remind us of the danger of war to which a State is constantly exposed. This is why a complete military organisation is the best security for a country in these days of gigantic armaments. The re-establishment of our means of communication and the formation of reserve forces are the objects to which patriotic prudence should tend—objects which are not unfortunately yet reached. It is certain, however, that we have obtained since 1870, despite difficulties of a political, financial, administrative, and military order, the required elements for our national defence. That dreadful war, by which were torn from us territories which Germany has not yet assimilated, was perhaps so far beneficial to France as to warn her of the dangers of an adventurous policy.   While it has inflicted upon us a loss in money of so many hundreds of millions, and has necessitated a complete renewal of our whole system of government, it has at all events been a terrible lesson for all governments, and especially for France.

CHAPTER VII.

WHEN the Isthmus of Suez was made we were merely realising the aspirations of the early masters of Egypt, for, according to the Arab historians, the Pharaoh who reigned in the time of Abraham had already conceived the idea of dividing the African isthmus, in honour of the visit of the patriarch and his wife Sarah, so as to establish communication by water between Egypt and Arabia.

We may ask, therefore, if it be true, as the old proverb has it, that there is nothing new under the sun, and that our ancestors discovered everything that required doing, and merely left to us, their descendants, the task of carrying out their designs? But even if this is so, we have no reason to be less proud, for is it not a glorious thing for us to be able to carry out the vast projects which they had conceived but were unable to realise, thus affirming the progress made by our race and age, in which all obstacles seem to have disappeared. The other day it was Suez, the isthmus of which was pierced, and

the writer of these lines may be pardoned for recall-
ing with pride how the year 1869 marked the realisa-
tion of a scheme which was desired by the Pharaohs
of the sixtieth century before Christ, of a work which
the men who built the Pyramids and drained Lake
Mœris were unable to accomplish.

A like work is now being undertaken upon the
American continent, upon the narrow neck of land
which divides North and South. The idea is not a
new one, for while America was discovered in 1472,
and Balboa ascertained the existence of the Pacific
Ocean in 1513, an attempt was made to unite the
two oceans in 1514. When the Spanish adventurers
ascertained that there was no natural passage between
the Atlantic and the Pacific, they conceived the idea
of cutting a canal through the spurs of the Cordilleras.
Just as it is certain that nature abhors difficulties and
encourages their overthrow, so it is certain that the
maritime trade of the globe ardently desires the
creation of a navigable zone which will enable it to
make the tour of the world, getting rid of the circuit
of Cape Horn as that of the Cape of Good Hope has
has already been got rid of.

The creation of a canal to unite the Atlantic and
the Pacific having given rise to much discussion, I
have thought it interesting to summarise what has
been said on the subject.

I.

The writings of the Spanish conquerors had, for more than two centuries, been consigned to the oblivion of the archives at Madrid, when the project of piercing the isthmus was revived. As soon as the impetus was given, there was a general outburst of enthusiasm among the hardy mariners and explorers who were eager to open a new route to the world's commerce. I should occupy too much space were I to quote all the names attached to this wonderful enterprise, but I cannot pass on without saluting the most famous among them, including Nelson, Childs, Lloyd, and our fellow-countryman Garella, and, above all, Thomé de Gamond, who was the first to propose the making of a tunnel between France and England, and he lived long enough to see it at all events begun.   There can be no higher reward for those who devote their lives to the pursuance of useful truths than to witness the commencement of the enterprise upon which their hearts are set.   From the year 1780 down to the present day a host of projects have been put forward for piercing the isthmus, some of them very carefully thought out and others purely fancy schemes.   But the last few years have produced more than the whole of the previous period.   The opening of the Suez Canal in 1869 produced a complete revolution in the commercial relations of the whole world, and I have no doubt that this event had a considerable influence upon the researches into the piercing of the American canal. For it is within the last fifteen years that so many bodies of explorers have gone out to investigate the nature of the work, and have come back loaded with valuable information calculated to throw light upon this intricate question. All honour to them for their zeal in assisting science to make this great step forward. At the same time, geographical studies which had been so much neglected in France, had, as a result of the war of 1870, which showed how necessary they were, again occupied public attention, and the learned societies which had inscribed geography in their programme commanded plenty of support.

Thus at the Antwerp International Congress, General Heine propounded the interoceanic scheme due to M. de Gogorza, and at the Paris Congress in 1875 the same subject occupied several sittings when I was in the chair. The information necessary for discussing the question in detail was not then forthcoming, and all that could be done was to express approval of the principle and convoke for a near date a special congress, or, it should rather be said, an international jury, to collect and collate all the necessary documents, and to form a definite opinion, after full deliberation, as to the technical and financial possibility of the work.

This resolution had the effect of giving a fresh impetus to the explorers and the authors of the scheme, all of whom were anxious to submit to the Congress complete and accurate plans. So that as soon as the proposed congress was announced, two companies were formed for making fresh expeditions, one of which visited Nicaragua, following the original route of Thomé de Gamond and Blanchet, while the other, under the conduct of General Türr, explored the more southern regions of Darien and Panama, marching in the steps of Garella, Lacharine, and Selfridge. The three years between 1875 and 1879 were fruitful in active researches and energetically conducted exploration. At the same date the expeditions set on foot by the United States were brought to a conclusion, and the able officers in command, Collins, Hull, Shufeldt, and especially Selfridge and Menocal, had left no part of the isthmus unexplored, while the documents which they brought back with them were calculated to facilitate the labours of the Congress very materially.

When the time arrived, and all the details relating to the recent expeditions were in my possession, I summoned the Congress, applying to all the savants, engineers, and sailors of the Old and New World, as well as to the chambers of commerce and the geographical societies, whom I asked to appoint delegates.

Few assemblies have included so many illustrious names as this great tribunal, which consisted of the leading representatives of science, politics, and industry. The first sitting was held on the 15th of May, 1879, at the meeting place of the Geographical Society, nearly every country being represented at the Congress. Mexico sent the engineer, F. de Garay, and China the mandarin Li-Shu-Chang. The United States were represented by Admiral Ammen, whose wide knowledge was of great service, Commander Selfridge, and the engineer, Menocal; while the countries of Europe had sent their leading geographers and engineers, such as Sir John Hawkshaw, and Sir John Stokes, Commander Cristoforo Negri, Signor de Gioià, the engineer Dirks, who cut the Amsterdam canal, and his colleague Conrad, President Ceresole, Colonel Coello, Dr. Broch, Admiral Likatcheff, Colonel Wouvermans, M. d'Hane Stenhuys, and many others whose names I ought perhaps to add, including all the most eminent scientific men in France. With an assembly thus composed, it was quite certain that the discussion would be frank, open, and luminous, and that the Congress would not separate until it had found a solution for the problem which was set before it.

The labours of this assembly will occupy an important place in history, and it will not, therefore, be thought that the space which I devote to the subject here is more than its importance deserves. In order to expedite its task the Congress was subdivided into five committees, each of which undertook to investigate one division of the very complex subject which

we had to discuss, and it is these commissions which
we have to thank for enabling us, by their scientific
labours and lucid discussions, to come to a speedy
conclusion.

The first, presided over by M. Levasseur, was
a statistical one, its task being to estimate the
probable traffic of the canal—that is to say, to go
through the customs' returns of all the ports of Europe
and America, and see what tonnage would in all pro-
bability pass through the canal.   I had had an opportunity of saying that the best course for the Panama,
as it had been for the Suez Canal, would be to prose-
cute the work by means of public money, and ask for
nothing from any of the governments, leaving the
enterprise its purely industrial character, and avoiding
anything like dabbling in politics.   The question,
therefore, was to know whether the capital invested
would obtain a sufficient return by the traffic passing
through the canal.   This was what the first commission had to calculate.

The second commission supplemented the work of
the first, and was called the Economic Commission.
After having calculated how many tons of merchandise would pass through the interoceanic canal, it
remained to be seen what income the traffic would
yield, and calculate, therefore, what tariff could be
charged vessels passing through.   Then it was necessary to estimate what would be the consequence of
the cutting of the American isthmus, what influence

the canal would have upon the trade and industry of each nation, and what new markets it would open to the trade of the whole world. The second commission, for which M. Simonin acted as reporter, was charged with the examination of the economical and financial results of the enterprise. The province of the third section was a more technical one, and it was composed of sailors, who discussed the influence of the canal upon shipbuilding, elucidated the regime of the winds and currents near the various canal routes submitted to the consideration of the jury, and pointed out under what conditions the safety and facility of the passage through the canal could be secured. This commission made an estimate of the speed of the vessels in pro-portion to the draught of water, and gave its opinions as to the effect of locks and tunnels in a canal intended to be used by the largest ships in existence.

The fourth commission was appointed to report upon the different routes for the canal submitted to the congress by their respective authors. Differing in this respect from the other sections, its functions were of a more general kind, as it had to discuss each project from an engineering point of view, to indicate the advantages and drawbacks of each, and fix what each would cost, both for construction and annual maintenance. The fifth commission was known as that of ways and means, and its duty was to complete, by entering into more details as to figures, the work of the second commission, and to name definitely the tariff which it would be desirable to charge, having regard to the probable earnings of the canal and the capital employed in making and working it.

The main object which we kept in view when forming these commissions was to draft as far as possible the most competent men into each of them. Thus the economists and geographers were placed in the two first sections, the naval men in the third, the engineers in the fourth, and the financiers in the fifth. They were all requested to be very reserved in their appreciations, and only to offer an opinion after the most careful scrutiny, so that the public might rest assured that there had not been the slightest tendency to take too optimist or enthusiastic a view of the undertaking.

The general results of the discussion are preserved in the reports of the public sittings, and more especially in the striking reports of the various commissions, which will remain an imperishable record of the history of the American Canal, and which must be read in detail in order to appreciate the lucid and learned information which they placed before the Congress. The most prejudiced will be constrained to admire the laborious efforts which enabled a hundred men, ardent in the pursuit of science, to place such a mass of evidence before the Congress during its brief session.

I propose to briefly review their labours, first of all examining the general considerations which were submitted to the international jury, and received its approval.

II.

The base of the problem to be solved was, as I have already said, the maritime traffic which it was neces- cessary to attract.

In the Statistical Commission, the principal repre- sentatives of the American States and the adminis- trators of the great maritime companies met under the presidency of Signor Mendès Leal. They first proceeded to examine the results of the working of the Suez Canal, which had then been open for ten years, and they asked for a report on this subject from M. Fontane, the Secretary-General of the Suez Canal Company, whose report made a deep impres- sion upon the Congress. M. Fontane proved, figures in hand, that an annual traffic of six million tons was only possible in a canal through which fifty ships could pass in the twenty-four hours. " This was why it was necessary," added M. Fontane, "in making the Suez Canal to adopt the system of a canal on one level without locks or drawbacks, to the exclusion of several very ingenious and bold plans presented by engineers of great repute." These views, which were the outcome of long and well-grounded experience, could not but have a marked effect on the minds of the members of the Assembly in respect to the choice which they had to make among the various systems submitted to them.

After having laid down this first and very important consideration, the Statistical Commission pursued
their task and prepared a voluminous report, the work
of M. Levasseur, whose scientific authority was a sure
guarantee against his giving reins to his fancy. The
plan which he adopted was proof against all criticism,
as he first sought to determine, by an examination of
the official returns of all the States, what tonnage
would take the route of the interoceanic canal. After
long and careful calculation, based upon the returns
for 1876, he estimated this traffic at £72,000,000, or
4,830,000 tons of merchandise. Taking into account
the annual increase in commerce, which for the years
1860-1876 was six per cent., he arrived at the conclusion that, with a much slower increase, the tonnage
would reach 7,249,000 tons by the time of the probable opening of the canal in 1890. This was the
minimum traffic of the canal as estimated by the
commission, and these figures are in no way surprising when the Pacific railway carries more than a
million tons, while the trade of Cuba exceeds 2,000,000
tons, and California alone produces 1,200,000 tons of
grain. Our figures are well within the mark, I am
sure, and they do not include, moreover, the transport of passengers, nor the large and small coasting
trade, which, at present quite insignificant, will develop with surprising rapidity in the Gulf of Mexico
and the West Indies.

The above-mentioned tonnage will show what an

important influence upon the history of the globe this new route will have. The labours of the second commission, presided over by Mr. Nathan Appleton, of Boston, completed this first report by showing what new markets would be opened, what new traffic would be created, and what advantages the traffic already in existence would derive from the cutting of the American isthmus. M. Simonin, the reporter of the commission, summed up these advantages in a very able report, which shows the distances that would be saved to navigators. From France and England, that is to say, from Liverpool, Havre, Nantes, and Bordeaux, the distance to San Francisco, round Cape Horn, is 5,000 leagues, whereas by Panama it would be only 1,500. For Valparaiso the distance would be reduced from 3,000 to 2,000 leagues. The saving in time for sailing vessels would be sixty days to San Francisco and thirty to Valparaiso. To this must be added the fact that steamers and sailing vessels alike would avoid the dangerous passage round Cape Horn. Thus the distance and the time in going from one part of the globe to the other would be materially shortened, and there would be such a reduction in the rates of assurance and freight that maritime intercourse would soon double itself, and that many markets now closed to European commerce would be opened, and provide it with fresh openings for import and export trade.

The New World will send us its woods, its indigo,

its coffee, its rice, its sugar, its india-rubber, and much of the mineral wealth which at present is only partially developed. Produce which at the present rate for freight is not readily carried, such as corn and fruit, will then be easy of export; and as produce is only exchanged for produce, the industry of Europe, receiving a fresh impetus, will send its manufactured articles all over the American continent.

The task of the Commission of Navigation, much shorter and more technical than that of the two first, was presided over by Dr. Broch, a former minister of the navy in Norway. It comprised several distinguished naval officers, such as MM. de Togorès, Linden, and de Marivault, and the heads of several great French and foreign shipping-houses. The report of its investigations, drawn up by M. Spément, a director of the Suez company, reviewed the probable influence which the cutting of the Panama Canal would have upon the transformation of shipping. He considered that the opening of the canal would favour sailing vessels even more than steamers, owing to the advantages derived by the former from the permanency of trade winds in the Gulf of Mexico. Speaking from another point of view, he recalled the fact that among the many schemes proposed, some involved the making of a tunnel, others that of locks. "As regards the tunnel," concluded the report, "the vessels would have to go through with their mainmasts up, and as the largest vessels, such as the *France* and the *Annamite*,

have very high masts, they would require an altitude
of nearly a hundred feet above the level of the water.
With regard to locks, they must be sufficiently numerous to admit of fifty vessels going through in a day.
This is the total which has been reached at Suez, and
there is no reason why it should not be equalled, and
even exceeded, by the Panama Canal. It would be
necessary, therefore, to have double locks, side by
side, one for vessels going west and the other for
vessels going east, and the construction of these would
entail special arrangements. In conclusion, therefore, I would say that a canal with locks ought only
to be accepted if a canal on the level is proved to be
impossible. So with regard to the tunnel, which
should only be adopted if it is found that, owing to
technical difficulties or excessive cost, the canal cannot be made without one."

III.

Thus far I have been explaining how three of the
commissions, without taking into account questions of
places, persons, or special schemes, treated the general
and theoretical part of the subject. To them it was a
matter of indifference whether the canal was by Thuyra
or the Bayano, by Nicaragua or Panama. In either case
the traffic would be the same, and the nations of the
east and of the west would derive the same advantages
from the making of the canal. The technical commission had quite an opposite task to perform, having to go

closely into the details of the subject, taking one after
another the numerous projects presented to the conference by their authors, to study them in detail so as
to bring out their commercial or technical advantages,
as well as to indicate their drawbacks and cost.   This
first work achieved, the technical commission had at
its command the necessary elements for comparing all
the projects, and selecting the one which it would
advise the Congress, at its plenary sitting, to adopt.
M. Daubrée, member of the French Institute, was
president, and Voisin Bey, formerly director of the
works of the Suez Canal, reporter.   The commission
comprised the most eminent specialists of all nations,
and it is quite certain that a decision ratified by the
names of Messrs. Hawkshaw, Dirks, Pascal, de
Fourcy, Favre, Couvreux, Lavalley, and Ruelle, who
carried as much moral as they did scientific weight,
would be beyond the reach of criticism.   Who better
than the creator of the Amsterdam Canal could treat
of the question of large locks ?   Who better than
the lamented constructor of the St. Gothard Tunnel
could discuss the question of the immense tunnel in
Panama, and the difficulties which would be entailed
in making it ?   Who more competent than Messrs.
Lavalley and Couvreux to speak of the cost of dredging and of excavating, both on dry land and under
water ?   Then, again, all the engineers who assisted
me at Suez had assuredly acquired the experience
necessary for settling the questions raised by the

examination of the various American projects for the canal.

The authors of all these projects appeared before the commission—viz., Messrs. Ammen, Menocal, Selfridge, de Garay, Blanchet, Belly, Wyse, Reclus, Mainfroi, and de Puydt—and expounded their plans, and met the objections which were advanced. This first operation, which occupied several long and interesting sittings, having been completed, the discussion began. Two important sub-committees were formed, one, which consisted of MM. de Fourcy, Voisin Bey, and five other members, being instructed to appreciate from a technical point of view, the character of the various routes; while the other, upon which MM. Ruelle, Favre, Lavalley, Couvreux, and Cotard sat, undertook to make an estimate of the cost of each plan, and to fix the probable earnings of it, based upon an identical scale of prices for each kind of work. It was between the reports drawn up by these two commissions that the Congress as a whole would be called upon to decide, and by making a summary of their investigations I shall best be able to give my readers an idea of the various schemes submitted to the opinion of the jury.

In order to explain them properly, I must say a few words as to the geography of the American isthmus, which extends a distance of 1,437 miles from the north-west to the south-east. Only the coasts and the banks of some of the principal rivers are inhabited, the interior of the country being so scantily peopled that the total population is only three millions, while France, covering the same area, has a population seven or eight times as large. There are next to no roads, and what few exist are very badly kept. Excepting these, the only means of communication are the rivers, and many of these are very difficult to navigate, as they are intersected by rapids, which the Indian avoids by carrying his canoe overland. The climate is a very torrid one, while it often rains for six months in the year, the annual rainfall at Panama exceeding ten feet. It is not surprising that, with such a high temperature and so heavy a rainfall, the vegetation develops with wonderful rapidity. Thus the organic life of the isthmus is very exuberant, and the virgin forests, with their gigantic cactus and cocoa trees, and their undergrowth, athwart which the native cuts a path with his axe or knife, form an inextricable network. It would almost seem as if all the venomous inmates of Noah's Ark had been emptied here, the country swarming with serpents whose bite is fatal, monstrous spiders, scorpions, and jaguars ; but, upon the other hand, it lends itself admirably to cultivation and industry, by means of which it would soon be completely transformed.

The ground is mountainous, the chain of the Andes rising to a height of over 13,000 feet, and presenting a striking contrast of volcanoes and of summits capped with snow. This is the land in which the canal is about to be cut; it is upon this wide cause-way, which separates North and South America, that the weak point in the armour has been found to effect a breach between the two oceans.

Let us begin with the north and go southward, following the report of the sub-committee. We come first to the isthmuses of Tehuantepec and Honduras; next to Nicaragua, then to Panama, San Blas, and Darien, each of these passages corresponding to one or more schemes for a canal, either on the level or with locks.

Señor de Garay, the Mexican delegate, dwelt with great force and sincerity upon the advantages offered by Tehuantepec for the tracing of the canal, but he met with little support. His scheme entailed a canal 150 miles long, with a maximum altitude of 975 feet above the level of the sea, to reach which 60 locks upon each slope would have been required. The cost of constructing these 120 locks and the fact that vessels would have been twelve days passing through the canal led to the immediate rejection of this project.

Seven or eight engineers, among them Messrs. Blanchet, Lull, and Menocal, brought forward plans for making the canal by way of Nicaragua. The geographical position of Nicaragua is, as a matter of fact, a very favourable one for the purpose, as in the centre of the isthmus a fine lake, 110 miles long by 35 broad, occupies the plateau which is 125 feet above

the level of the Atlantic. This lake receives the
waters of some forty streams, and flows into the
Atlantic through that noble river, the San Juan.
Unfortunately this stream is intersected by several
cataracts which render navigation impossible. One
of the worst of these cataracts is human handiwork ;
for the inhabitants of the colony, to protect themselves
from the fillibusters who ravaged the West Indies in
the seventeenth century, obstructed the course of the
San Juan by sinking vessels in it with trunks of trees
and large masses of rock. The water being driven
back found a fresh outlet at the side of the San Juan,
and this outlet, now known as the Rio Colorado, has
never been stopped. In order to improve the navigation of the San Juan it would be necessary to
canalize it by means of seven or eight locks, and
to regulate its course by an immense embankment
twenty-eight miles long upon the other slope. It
would further be necessary to intersect the Rivas
with a deep trench, make seven more locks, and
create at the two ends of the canal Greytown and
Brito, harbours upon coasts which are very unsuited
for the purpose. The partizans of these projects
urged in their favour the superiority of the climate,
the abundance of materials in the country, and the
relative density of the population ; and it was very
clear that if the canal was to be one with locks, this
would have been the best of them. The total length
of the canal, including the 55 miles of the upper lake,

would have been 182½ miles, and the time occupied in going through it four days and a-half.

The Americans, through the mouthpiece of Admiral Ammen, were very much in favour of this project, which was admirably conceived and propounded by one of their engineers (Menocal). A French engineer, M. Blanchet, proposed to amend it by prolonging the summit-level of the Valley of San Juan, and by substituting for the seven locks which formed part of the American scheme a large work with 105 feet difference of level, which had been designed by MM. Ponchet and Sauterean, and which one of our most distinguished constructors, M. Eiffel,* was to have carried out. The gates of this lock were to have weighed nearly 1,000 tons, and to have been 23 feet thick.

Two officers in the French navy, Messrs. Wyse and Reclus, who had explored the country with great perseverance, presented a scheme for cutting a canal on the level through the Isthmus of Panama, and before they had proceeded far with the explanation of their scheme, it was clear that they had made a deep impression upon the members of the Commission, and that herein lay the solution of the problem. If objections were raised at first, this was rather, it seemed, with the view of disposing of them, so as to

---

* Note of the Translator.—M. Eiffel is now erecting the iron tower, 1,000 feet high, which is to be one of the features of the Paris Exhibition in 1889.

be free to consider, with perfect freedom of mind, all
the advantages which the project presented. The
Wyse canal was to follow the thalweg of the river
Chagres, pass under the Cordillera by means of an
immense tunnel, and reach the Pacific slope by the
valley of Rio Grande. In the course of the discussion
the authors of this scheme, in obedience to the advice
given them, agreed to substitute for the tunnel a
deep cutting in the mountain, and the Mexicans, it
may be added, have set the example in this respect,
the cutting at Desague being 220 feet through, while
that of Panama will not exceed 290 feet. Two objections had struck the Technical Commission, and
it was, I think, very striking evidence of the advantages which the Panama project possessed in the eyes
of the experienced engineers sitting upon it, that it
was they who urged the authors of the project to
overcome their objections.

The first of these objections bore upon the sudden
risings of the Chagres River. This river rises so
rapidly that it has been known to rise more than
twenty-five feet in a single night. The question was
how to get rid of the waters, the irruption of which
would have been dangerous in the making and working of the canal. M. Wyse first proposed to form a
vast reservoir of the overflow of these waters, in
immense excavations which would admit of an outflow
of over 330 cubic yards a second. But this did not
satisfy the Commission, which urged that it was no

trifling affair to create an artificial lake of this kind, and to maintain such a mass of water suspended 100 feet above the canal. Why not free the canal entirely and make a separate bed for the river ? This was the solution upon which the authors of the scheme eventually agreed, at the instant advice of the Commission.

The second objection was that the Pacific tide is 19½ feet at Panama, while the Atlantic tide at Colon is only two feet. This would cause currents running four or five knots an hour in the canal, and create a danger to navigation. The remedy for this will be to create a tidal gate at Panama, and place at the entrance to the canal a waiting basin, where ships can pay the customs and transit dues while waiting for a suitable hour to enter the canal.

If to this we add that the Panama Canal passes within half-a-mile of the railway, that the latter will be most useful for bringing labourers and materials to the works, and that the length of time occupied in going through the 47½ miles of canal will be only thirty-six hours, the words of the sub-Commission need no further justification:—" The Panama canal on the level technically presents itself under the most satisfactory conditions, and ensures every facility, as it gives every security, for the transit of vessels from one sea to another."

I must say a word about the San Blas Canal. Advocated by Messrs. Appleton and Ralley, this canal had in its favour the fact of its being shorter than any of the others, its length being only 33 miles, but of this nine miles were tunnel, while the river Bayano had to be diverted from its course, so that the Technical Commission felt bound to reject it.

Upon the other hand, the Commission examined with the utmost care and interest the remarkable researches of an officer of the American navy, whose name I have already mentioned, Commander Selfridge. The Selfridge scheme followed the Darien Isthmus and the Atrato River, which it was to canalize for a distance of 150 miles, and it then made a sharp bend southward, and reached the bay of Chiri-Chiri by a cutting and a tunnel two and a-half miles long. But the question was, whether this Atrato River, the mouth of which formed a vast and marshy delta, could be so deepened as to ensure over twenty-five feet of water at its bar, and, if so, how this depth of water was to be maintained? Then, again, it was difficult to see how the risings of the Atrato were to be foreseen, and their effects alleviated, so that the Commission felt compelled to reject Commander Selfridge's scheme.

The Commission also examined, just as it was about to break up, a scheme which its author, M. de Puydt, produced without any documentary evidence to back it up, and which proposed to cut the canal through Darien, from Puerto Escondido to Thuyra. The watershed by this route was the pass of Tanela Paya, the slope of which, according to M. de Puydt, is only

150 feet, so that the canal could have been on the level. The author's figures were, however, given without anything to support them, and were directly contradicted by other explorers; and it was only in order to show its absolute impartiality that the Commission thought right to examine his project.

When all was done, two projects alone were before the Commission: one for making the canal through Nicaragua, the other through Panama.

The first, which was the less costly, as it was estimated to involve an expenditure of £32,000,000, while the latter was to exceed £40,000,000, was at the same time more limited in its scope, and longer in point of distance and time. The objections to it were its sixteen locks, its reaches, which the vegetation of the tropics would cover with terrible rapidity, its works of art, which the slightest shock of earthquake might destroy, and the care and deliberation which the handling of so much delicate apparatus would entail. There was nothing of this kind to apprehend with the Panama Canal, which was a fourth shorter than the other in point of distance and a third in point of time, while it did not entail any works of art, or set any limit upon the number of ships which could pass through it in the twenty-four hours. This was surely sufficient to justify the decision of the Technical Commission.

Upon the proposal of the engineers of the Suez Canal, the Commission decided by a large majority against

the system of locks, and declared strongly in favour of
an open canal on the level, the feasibility of which
seemed quite clear if the Colon-Panama line was
followed.

But compelled by its mission to make a choice between the various schemes submitted to it, the
Commission was nevertheless desirous of testifying
to how carefully most of them had been thought
out, and to the talent of their authors. "More
especially," to borrow the exact words of the
report, "to the eminent American engineers and
explorers whose admirable researches will remain
as a monument in the history of this gigantic
undertaking." The Technical Commission also
pointed out how the canal should be made, that the
curves should not be under 1¼ miles, that it should
be 72 feet wide and 28 feet deep, and that there
should be only one canal as at Suez, but with numerous sidings to admit of ships passing one another,
all the details of execution having been carefully foreseen and discussed at this Congress, from which those
who are now making the canal cannot fail to derive
most useful lessons.

When the Technical Commission had terminated
its works and fixed the figures at which it estimated
the cost of making and maintaining the canal, and
when, upon the other hand, the Economic Commission had laid before the Congress all the elements
required for calculating the transit, the fifth section,

that of Ways and Means, was able in turn to accomplish its part with these data for its guide. M. Céré-
sole, the ex-president of the Swiss Confederation, was
the president, and M. Chanel, the delegate of Mar-
tinique, was reporter, the judgment of the section
being : "We are convinced that the sum of the elements of transit, already amply sufficient to defray
the cost of the canal, is destined, as the work develops,
to expand to an incalculable extent."

The report went to show by what series of calcula-
tions the Commission had been led to fix the transit
dues at fifteen francs (12s.) per ton.

Going on to calculate the cost of construction, the
payment of interest, the annual cost of working and
of maintenance, and deducting the participations
reserved by the Act of Concession granted by the
Government of Colombia, the reporter, and with him
the Commission, estimated the net annual profit of the
canal at £1,680,000. And, finally, "to guard against
the risks and chances of the unknown," the Commission expressed their hope "that, even at the cost of
more time and money, the canal might be made without locks or tunnels."

It is a very remarkable fact that the five Commissions of the Congress should, without any pre-arranged
understanding, have expressed the same wish and
displayed their aversion for a canal with locks. But
this agreement of views simplified the remainder of
the proceedings. When, according to the mode of

procedure agreed upon, the five Commissions had
communicated the result of their deliberations, all that
the bureau of the Assembly had to do was to co-
ordinate these conclusions, in order to draw up and
submit to the Congress the resolution which was to
be the outcome of them.

IV.

" The Congress is of opinion that the cutting of an
interoceanic canal with one level, so desirable in the
interests of trade and navigation, is possible, and that
this maritime canal, in order to give the indispensable facilities of access and use which a passage of
this kind must be supposed to give, should go from
the Gulf of Limon to the Bay of Panama."

Such was the form of resolution adopted by the
bureau and reinforced by the presidents, secretaries,
and reporters of the five Commissions. It was put to
the vote on May 29th, 1879, and out of ninety-eight
members present seventy-eight voted in its favour
and eight against, the twelve others abstaining. Such
was the majority which declared in favour of the
canal, recompensing the bold and persevering efforts
of our compatriots, Wyse and Reclus. If we examine
the nature of the voting, we may see that there was
something like unanimity, for among those who voted
against the resolution, or did not vote at all, were the
representatives of the Northern States of Central
America, whose local sentiments were enlisted in

favour of the Nicaragua Canal. These included the able constructor, who had been selected to make the large lock of Nicaragua, and the president of the association for cutting that canal, yet both of them cheered the announcement of the vote.

It is characteristic that among those who gave in their adhesion to the scheme were the Dutch engineer, who had constructed the Amsterdam locks, Commander Selfridge, who explicitly declared that his countrymen would accept the decision of the Congress without any reserve or afterthought, the engineers of the Suez Canal, and many others whose statements were enthusiastically cheered by the public.

The course which the Congress approved was that which had been traced by Lloyd, Totten, Garella, Wyse, and Reclus. It strikes the Isthmus at the ninth parallel, between the Bay of Limon upon the Atlantic and the Gulf of Panama on the Pacific. It is not half as long as the Suez Canal, being only 45½ miles long instead of 101; it has two excellent ports at each end, is close to two good towns and to a district thickly inhabited, and has a railway in full working order. Such is the country which the canal will traverse, transform, and enrich.

Carrying my mind back a few years, I cannot but remember how many people—including several eminent men, too—formerly treated the Suez enterprise as impracticable. They said that it was madness to try and create a port in the Gulf of Pelusium, to traverse the mud of Lake Mensaleh and the entrance to El-Guisr, to pass through the sand banks of the desert, and form workshops twenty-five leagues away from any village, in a land which had no inhabitants, no water, no roads, to fill up the basin of the Bitter Lakes, and to prevent the sand from silting up in the canal.

Yet all that was accomplished, at what a cost in labour and perseverance I well know; and I maintain that the Panama will be easier to make, easier to complete, and easier to keep up than the Suez Canal.

Nothing has occurred since 1879 to alter the aspect of affairs from a material point of view, and it is not for me to discuss here the motives of the eleventh hour opposition, raised in order to prevent the success of the subscription which, after the vote of the Congress, it seemed to me opportune to open.

I will merely repeat what I said at the Académie des Sciences :—

"The line from Colon to Panama can easily, according to the latest data of science, be utilised for the cutting of a salt-water canal on one level in preference to any other route necessitating locks fed with fresh water. The experience of the Suez Canal has shown that, in order to ensure a considerable amount of transit navigation, you must have a maritime canal as free as a natural Bosphorus, and not a river canal, subject to stoppages more or less lengthy, and only fit for internal navigation."

To this I may add what I said in a circular which was published at the time :—

"The arguments of the opposition may be summed up as follows : Upon the one hand the expenses have been exaggerated and the receipts under-estimated, in order to show that if the idea of opening a new maritime route to trade and to civilisation is good in itself, the enterprise is financially bad. Upon the other hand, an effort has been made to create uneasiness by representing the United States of North America to be hostile to the scheme. The first argument has been met by the able contractor who removed the bed of El-Guisr, at the entrance to the Suez Canal. M. Couvreux and his associates, who are responsible for the regulating of the course of the Danube, and for enlarging the ports of Antwerp, are at this moment engaged in investigating, at their own expense, the work required for making the new canal. They have determined to undertake to execute the work either by contract or for a royalty, as I may prefer, and thus to leave no doubt as to the real amount of the expenses. With regard to the second objection, I shall solve that myself by an early voyage to America."*

Heer Dirks, the Dutch engineer who cut the canal which connects Amsterdam with the sea, has expressed his surprise at what he terms "the malignant attacks and anonymous notes inserted in various

* Note of the Translator.—This circular was issued several years ago—in fact, before the work of cutting the canal had been begun.

papers," and adds : "All anonymous attacks are worthless and condemn themselves, whereas a frank and open opposition is of service to those who deserve it."

I may add that I have never been alarmed by the obstacles thrown in the path of a great enterprise, nor by the delays which discussion and contradictory arguments entail, my experience having taught me that what is accomplished too quickly has no deep roots, and that "time hallows only that which he has himself made."

CHAPTER VIII.

THE expansive force of steam has long been known, but its perfected use is of contemporary application. In 1830, the French fleet which took part in the Algerian expedition included 500 sailing vessels of an average burden of 500 tons for a body of 30,000 men, and one steamer, the *Sphinx*, of 160 tons.

In 1880, the number of vessels which went through the Suez Canal, carrying 100,000 soldiers and as many civilians, was 2,025, and they were of 4,344,465 tons burden, or 2,145 tons each.

After centuries of war and destruction, steam and electricity seem likely to open an era of unlimited progress, by multiplying the means of pacific communications between the peoples of the earth. Let us go back for a moment to the origin of the invention of steam power and its various applications.

I.

England, as regards maritime navigation, and the United States as regards fluvial navigation, having

anticipated France in the perfected use of the locomotive and the steamer, we are inclined to forget
that the real invention of machinery as applied to
navigation is due to two Frenchmen, Denis Papin
and Claude Jouffroy.

Aristotle and Seneca seem to have been the first
to suspect the expansive force of steam, for they
attributed earthquakes to the transformation of water
into steam by the subterranean fires, a theory which
quite fits in with the present teachings of science.
Seneca, more explicit still than Aristotle, compares
the volcanoes to boiling water running out over the
sides of a vessel under the action of fire. Four hundred years after Aristotle, Seneca, in chapter vi. of
his *Natural Questions,* wrote :—

"Certain philosophers, while attributing earthquakes to fire, also ascribe to the latter another
action. Fire, they say, when lighted in several places
at once, carries with it abundant vapours, which,
having at first no outlet, communicate to the air
with which they mingle a great expansive force.
If the air, thus charged, acts with great energy,
it breaks down all obstacles ; if it is more moderate in its power, it merely causes the ground to
quake.

"We see water boiling upon the hearth, and we
may be sure that if this limited phenomenon takes
place inside a vessel, it assumes tremendous proportions when vast fires are acting upon vast masses of

water. These vaporised waters overcome all obstacles and overturn everything upon their passage."

Hero of Alexandria, surnamed the Ancient, who lived about 200 B.C., composed several works on physics, only three of which are extant. The reacting engine is defined and represented in the treatise entitled, *Spiritalia, seu Pneumatica.*

Description of the Eolipylus (Gate of Eolus).

BY HERO OF ALEXANDRIA.

This, after the fragment translated into French by M. Egger, is described as follows :—

" A vessel being heated from underneath, a sphere is made to turn upon its pivot. Or else a vessel containing water, and with a lid over the orifice. To this lid should be adjusted a tube bent so that one end of it may be embedded in the side of a hollow sphere. Opposite the end of the tube, and following the diameter of the sphere, should be a pivot rising over the lid; let the sphere be fitted with two small bent ajutages fixed to its side, according to a corresponding diameter, and bent the reverse way the one from the other. Suppose for a moment the elbows of the ajutages upon the vertical plane. Thereupon, the vase being heated, the vapour, ascending into the sphere through the tube, will escape through the ajutages of the elbows above the cover, and will make the sphere move upon its axis, as is done with persons asleep."

It is probable that Hero of Alexandria imitated the
procedure of the priests of ancient Egypt, who, it is
said, caused inanimate objects to move, or doors to
open and shut at their bidding, by means of tubes let
into the passages.   Many tourists have seen the
colossal statue of Memnon, which emitted sounds
when struck by the sun's rays in the burning plain
of Thebes.   The escape of the vapour caused by the
damp which had found its way in through the inter-
stices, and had been produced by the radiation of the
cold at night as well as by the abundant morning
dew, quite explains this phenomenon.   At the base of
the monument may still be read inscriptions in prose
and in verse testifying to the wonder of the Greek
travellers.

There is now in the head of the Colossus a fissure
through which an Arab, for a small fee, will, after
having managed to climb up, pass his arm and
produce a metallic sound, by striking the hollow space
inside with a stone.

By way of a connecting link between the Greek
engineer Hero and modern authors, we have the
following passage from Rabelais, which Littré quotes
in his Dictionary :—

"Eolipylus, gate of Eolus.   It is a closed instru-
ment with an opening through which, if you place
water and put it near the fire, you will see wind
constantly pouring forth."—(Rabelais, notes on Book
4, chapter xliv.)

The Spanish archives of Simancas contain the following document :—

"Blasco de Garay, sea captain, submitted, in 1543, to the Emperor and King Charles V., a machine for propelling ships and large boats, even in calm weather, without oars or sails. Despite the obstacles and difficulties which the project encountered, the Emperor ordered trial to be made of it in the port of Barcelona, which trial took place on the 17th of June, in the said year 1543.

" Garay would not entirely divulge his discovery. But it was observed at the time of the trial that his machine consisted of a large cauldron of boiling water and of revolving wheels attached to both ends of the vessel.

" An experiment was made on a 200-ton vessel called the *Trinity*—Captain, Peter de Scarzo—which had just arrived fron Colibra with a cargo of wheat. By order of Charles V., Don Henry of Toledo, the Governor Don Peter of Cardona, the Treasurer Ravajo, the Vice-Chancellor, and the High Steward of Catalonia assisted at these experiments, and in their reports to the Emperor they spoke approvingly of the invention. The Treasurer Ravajo, however, who was opposed to the project, said that the vessel would not travel more than two leagues in three hours, that the machinery was very complicated and expensive, and that there was a great danger of the boiler bursting. The others affirmed that the vessel put about as readily as a galley manœuvred in the ordinary way, and went at least one league an hour. After the trial Garay took away the whole of the machine, leaving only the wood-work in the Barcelona arsenal. In spite of the opposition of Ravajo, the invention of Garay was approved of, and but for the expedition in which Charles V. was engaged standing in the way, he would no doubt have favoured its adoption. As it was, the Emperor raised him a step, made him a present of 200,000 maravedis, and ordered the Treasury to pay all his expenses."

Arago, referring to this in his lecture to the students of the Polytechnic School, said, "As Garay would not show his machine to anyone, not even to the commissioners appointed by the Emperor, it is of course impossible, after the lapse of three centuries, to say of what it consisted. The document, exhumed from the archives of Simancas, in 1825, must be put on one side, first, because it was never printed; second, because there is no evidence that the motive power of the Barcelona boat was steam; and thirdly, because if a Garay locomotive ever existed, it was to all appearances the Eolipylus described in the works of Hero of Alexandria."

Salomon de Caus is the author of a work entitled *Les Raisons des forces mouvantes avec diverses machines tant utiles que plaisantes.* This work appeared at Frankfort in 1615, and it contains the following theorem (No. 5) thus set forth: "Water will rise by means of fire higher than its own level." The Marquis of Worcester, whom the English regard as the real inventor of the fire-engine, lived in the reign of the Stuarts, and having lost his immense fortune during the revolutions of those times, he was cast into prison, but escaped to France. Returning to England, he was detected and shut up in the Tower of London. It is said that Worcester's idea as to the impulse which steam could give originated in his remarking how the lid of the saucepan in which his food was being cooked was suddenly lifted up. A second edition of Salomon de Caus's book had appeared in France while he was residing there. Worcester's apparatus is thus described in his book entitled *A Century of Inventions:*—

"I have discovered an admirable and very powerful means of raising water by means of fire, not by suction, for then, as the philosophers say, one would be limited *intra spheram activitatis*, as suction only operates for a given distance. But there is no limit to my means if the vessel is strong enough. By way of trying it, I took a whole cannon, the mouth of which had burst, and three parts filling it with water, I closed the end which had burst and the touch-hole with screws. I kept up a very strong fire inside, and in twenty-four hours the gun broke up with a loud report."

Denis Papin (1690-1695).—The machines of Salomon de Caus and the Marquis of Worcester were merely apparatus for raising water. This was the first object which Papin had in view with his engine, but at the same time he had quite seen that the up and down movement of the piston on the body of the pump could be applied to other uses. I may perhaps be permitted to quote in this connection a few extracts from a speech which I made at Blois on behalf of the Académie des Sciences, at the inauguration of Papin's statue on the 29th of August, 1880.

I said: "The great inventions destined to change the face of humanity rarely enter the domain of accomplished facts until they have passed through what may be regarded as a providential network of experiments, which may be isolated, but which are summed up and applied by the close researches of a man who is at once perspicacious and disinterested, who knows no guide but science, and who has no object but that of being useful to humanity, disregardful of the atmosphere of errors and prejudices amid which his discoveries are conceived and put in action.

"Denis Papin was one of these exceptional men. The following is the summary of his labours and discoveries:—

"1674-1709. Perfecting and modifying the pneumatic engine.

"1681. Apparatus known in the present day as Papin's digester, autoclave, etc. The guidance of steam. Safety valve.

"1685. Discovery of the principle of air-pressure syphons.

"1687. Discovery of atmospheric locomotion.

"1695. Fumivorous apparatus, or apparatus for the

combustion of smoke. Doubly exhausting stop cocks, of which Watt and Leopold have made one of the principal features in the high-pressure steam-engines, where the barrel might be used for other purposes. He also discovered a method for transforming the reciprocating motion into a rotary motion. Papin invented the first piston engine. He was the first to note that vapour of water affords a very simple means for obtaining a vacuum in the capacity of the barrel. He was the first to whom it occurred to combine in the same engine the action of the elastic force of steam with the power which, as he pointed out, this same vapour possesses of condensing itself as it cools."

Captain Savery, an Englishman, who lived at the end of the seventeenth century, made some inventions in the same line, which are referred to by Arago as under :—

"We have no proof that Salomon de Caus ever constructed his steam-engine. I might say the same of the Marquis of Worcester. Papin's engine in which the action of the steam and its condensation are successively brought into play was only executed in miniature and with a view to make an experimental trial of the exactitude of the principle upon which it was based. So that although there was nothing very new in Savery's steam-engines, it would be very unjust not to mention them, as they are really the first which were put into practical use. According to Salomon de Caus's plan the motive steam was to be engendered in the vessel containing the water and by means of this same water. In Savery's engine there were two separate chambers, one containing the water and the other, which may be called the boiler, the steam. This steam, when a sufficient quantity has been generated, finds its way to the upper part of the water chamber by a communicating tube which can be opened at will by means of a tap. It exercises a downward pressure upon the liquid surface, and forces it back into a vertically ascending tube, the lower orifice of which must always be beneath this surface, for otherwise the steam itself would escape.

"In Salomon de Caus's engine, as soon as the presence of the steam has produced its effect, a workman has to make good the water which has been driven out by means of an orifice in the upper part of the metallic sphere which opens and shuts at discretion. All that then remains to be done is to keep the fire going. In Savery's engine the water is let in, not by a workman, but by atmospheric pressure.

"In short, Savery sought to utilise steam for driving water into a vertical tube, but Salomon de Caus had done precisely the same thing eighty-three years before. Savery, again, effected the vacuum which brought about the suction by the cooling of the steam. This was a very important matter, but Denis Papin had long before drawn attention to it."

SUMMARY.

1615. Salomon de Caus was the first who conceived the idea of utilising the elastic force of vapour of water in the construction of an hydraulic pumping engine.

1690. Papin conceived the possibility of making a steam and piston engine. He was the first to combine in one and the same steam and piston engine the elastic force of vapour of water with the precipitating property which steam acquires through cold.

1705. Newcomen, Cawley, and Savery were the first to see that in order to effect a rapid precipitation of vapour of water, the injected water must find its way into the mass of steam in the shape of very small drops.

1769. Watt explained the immense advantages, from an economical point of view, obtained by substituting for the condensation which had hitherto been effected in the barrel of the engine condensation in a separate chamber. He was the first to point out the advantage which might be derived from the expansion of the vapour of water.

Chaillot's steam pump was made after his plans in the workshops of the brothers Perrier.

1783. Jouffroy, in the presence of thousands of spectators, made the first trial of a paddle-wheel steam-boat, which he had constructed himself, and which went up and down the river Saône, between

Lyons and the Ile Barbe. This steamer was 150 feet long by $14\frac{1}{2}$ feet in diameter, with a draught of rather over 3 feet of water, and a speed of two leagues an hour.

1801. The first locomotive high-pressure engines made by Messrs. Trewithiet and Vivian, Englishmen.

1807. Fulton applies steam navigation to the great American rivers.

II.

Papin must be considered the first inventor of the steam-engine and of the idea of applying it to navigation. But his first attempt could not be practically tested owing to the destruction of his machine by the populace before the experiment took place, and the glory of having executed the first steamer which ever navigated a stream belongs to Claude de Jouffroy. This young nobleman of the Franche-Comté belonged to a class which, especially in his neighbourhood, set but scant store by scientific studies. With a few exceptions, the country nobility had a horror of any kind of trade. The scientific tastes of Claude de Jouffroy, the singular aptitude with which nature had endowed him, were a source of annoyance to him at home. He was laughed at in the drawing-rooms of his neighbours and nicknamed "Jouffroy the Pump." Even at Court, where the report of his experiments had preceded him, people pointed him out to one another, and said: "Do you know this young man of the

Franche-Comté, who embarks steam engines upon rivers, this lunatic who would have us believe that he can marry fire and water?"

In order to escape from the yoke of the prejudices which surrounded him, Claude de Jouffroy determined to take service in the artillery, so that he might be able to utilise the experience which he had gained. But there was a great outcry at this, for the nobility at this period considered it derogatory to enter that branch of the service, leaving the artillery and engineers to the middle classes. Having been a page to the Dauphin's wife, and having entered at the age of twenty the Bourbon regiment as sub-lieutenant, he had a duel with his colonel. He was then exiled for two years to the island of St. Marguerite, opposite Cannes. During his enforced leisure, while watching the galleys and their oarsmen, he was struck by the drawbacks of this mode of navigation, and conceived the idea that the use of steam as a motive power might obviate it. When his exile was over, in 1775, he went to Paris, where the brothers Perrier had just founded a large establishment, and had imported from Birmingham one of Watt's engines, known in France as the "Pompe à feu de Chaillot."

Jouffroy met in Paris two men from his own district, soldiers like himself, the Comte d'Auxiron and the Marquis Ducrest, colonel in the Auvergne regiment, brother of Madame de Genlis, member of the Académie des Sciences, and author of a work on mechanics. Count d'Auxiron encouraged him strongly to persevere, and wrote to him from his deathbed, "Be of good cheer, my dear friend. You alone are right!"

Jouffroy, having no influence in Paris, went back to his own province, where, full of confidence in the future of his idea, left to his own resources, and having no guide save his own persevering studies, and no other workman than a village tinker, he succeeded, in 1776, in constructing a machine which he adapted to a boat. This first steamer was about forty-two feet long by seven feet broad, and the floating apparatus consisted in rods about eight feet in length, suspended upon each side of the forepart of the vessel, and having at their extremities chains fitted with movable two-feet wooden flaps. The chains described a radius of eight feet, and a lever fitted with a counterweight kept them in their place. A single Watt engine fixed in the centre of the boat set the articulated oars in motion. The construction of this apparatus, in a place where it was impossible to procure drilled cylinders, was a work of genius, courage, and patience ; and, despite its imperfections, the apparatus was superior to anything which had hitherto been proposed for navigating purposes. The boat was in use on the river Doubs, at Baume-les-Dames, between Montbéliard and Besançon, during the months of June and July.

Somewhere about 1780 Jouffroy came to Lyons, in the hope of obtaining the funds required for perfecting his invention, and while there he married Mdlle. Madeleine de Vallier, and fitted up a fresh apparatus in the smithy of the Messrs. Frèrejean.

The dimensions of this second boat were, as already stated, very much larger than those of the first, and in it he ascended the current of the Saône, from Lyons to the Ile Barbe, on July 15th, 1783, in the presence of a committee of savants and of thousands of spectators.

After repeating his experiments with unvarying success, Jouffroy entered into partnership with MM. de Follenay, Auxiron, and Vedel, with the view of founding a steam navigation company for the conveyance of passengers and goods, first of all upon the Saône, and afterwards upon the Rhône and the other navigable rivers of France. Another financial company offered to join him, upon condition that the founders would secure for it the privilege of working the enterprise for a period of thirty years.

This privilege was not secured, as appears from a letter which M. de Calonne wrote from Versailles on January 21st, 1784. The boat continued to ply on the Saône for sixteen months, and was then abandoned.

Jouffroy was completely ruined during the Revolution, but in 1815 he obtained a patent for invention and improvement, and built a boat named *Charles-Philippe*, after the Comte d'Artois, which was launched upon the Seine on April 20th, 1817, in the presence of the Comte d'Artois, his sons, the Paris municipal authorities, a great number of learned men, and a crowd of spectators. All promised well for the prosperity of the enterprise, when a rival company in turn obtained a patent, disputed Jouffroy's claim to priority, and brought from England a boat fitted with their engines. The competition in a mode of navigation against which prejudice was still so strong proved disastrous to both companies.

Jouffroy, whose faith in the future of steam navigation was not to be shaken, once more retired to his native district to get together the means for starting a fresh society, and, with the help of a few intelligent friends, he succeeded in forming a capital of £960, divided into twenty-four shares of £40 each. This small capital was spent in the construction of a steamer called the *Persévérant.* Upon July 8th, 1819, the partners agreed to constitute a capital of £8,000 for the construction of several steamers, so as to organise a regular service. The *Persévérant* plied for several months between Lyons and Chalons. Prejudice and conflicting interests prevented the creation of the required capital, not that anyone denied that this mode of transport was speedy, but they urged that navigation was impossible on the Rhône and full of obstacles on the Saône, owing to shallowness of the stream, and that the powerful Compagnie Générale des Transports would not stop at anything to put down competition. So great were the obstacles in the way of steam navigation at Lyons, even twelve years after it was prospering in America, and after Henry Bell had overcome the prejudices which marked its introduction upon the coasts of the United Kingdom.

In this same year (1819) Captain Moses Roger crossed the Atlantic, from New York to Liverpool, in a compound sailing and steam vessel of 380 tons.

Foreign capitalists gathered; even in France, the fruit of the labours upon which Jouffroy had for half a century concentrated all the resources of his genius and his fortune.

In the year following, Steel, an English builder, launched upon the Seine a steamer provided with an articulated oar or goose-foot, after the first system tried by Jouffroy. Two years later, an English company brought two iron steamers into France. In 1825, a compound English steamer made a voyage from Falmouth to Calcutta, and a Dutch boat of the same kind went from Amsterdam to the West Indies. From 1825 to 1830 nearly all the navigable rivers and ports of France used steam-boats.

The problem of the employment of steam for trans-atlantic voyages was definitely settled in 1830 by the passage of the Great Western (1,300 tons) from Bristol to New York, and by that of the Syrius (700 tons) from Cork to New York.

What, it may be asked, had become of Jouffroy while all this progress was being made? In 1829 he lost the wife whose goodness of heart and intelligence had consoled him during these forty-six years for all his disappointments, and, unable to endure the

solitude which her death inflicted on him, he liqui-
dated his retiring pension as captain in the army, and
got admitted to the Hôtel des Invalides, where he
died of cholera in 1832, at the age of eighty-one,
leaving to his children no other inheritance than the
example of the laborious life which his eldest son so
loyally followed.

FULTON.

At the close of the last century, a young American,
who had been at school while the War of Independence was in progress, came to study art, for which
he showed great aptitude, in France, although he had
no special genius for invention, he was endowed with
great readiness in the study of mechanical discoveries,
and with a perseverance which no rebuff could retire.

Of Irish origin, and born at Little Britain (Penn-
sylvania) in 1765 of parents who had emigrated in
a state of great poverty, Robert Fulton was first
apprenticed to a jeweller, and afterwards to a painter.
At twenty years of age he left America and passed
ten years in England, where he devoted himself
entirely to the study of mechanics, coming to Paris
in 1796. For five years he concentrated his attention
upon submarine navigation, and upon the means of
exploding at a given point boxes filled with gun-
powder, so as to blow up vessels on the water.

The French Government refusing to adopt this
invention, Fulton was about returning to America,

when he met Chancellor Livingston, then Ambassador of the United States in Paris, who was then studying the question of steam navigation in the company of an Englishman named Nisbett and the French engineer Brunel, who afterwards made the Thames Tunnel. Livingston undertook to find the necessary funds for establishing steam navigation in America, and Fulton, after making a study of the previous essays, decided to adopt the paddle-wheel. Experiments made on the Seine (August 9th, 1803), before a committee of the Académie des Sciences, proved a complete success, but Napoleon refused to let the question come before the Academy, for, as England at that period alone had large workshops for the construction of the machinery, she would have benefited by the invention long before France would be in a position to utilise it. Moreover, Fulton frequently stated that it was his intention to establish steam navigation upon the broad American rivers, and not on what he called the rivulets of France. A steam-engine ordered by Livingston and Fulton, unknown to Bolton and Watt, in 1804, was only ready in October, 1806, upon which date Fulton sailed for New York, and launched his boat on the East River. When his success in the States was placed beyond all question, the priority of his claim was disputed, and the worry of the lawsuit undoubtedly hastened his death, which occurred when he was only fifty, on February 24th, 1815. The legislature went into mourning for him for a month, but his family was left very badly off.

Fulton never questioned Claude de Jouffroy's priority in the practical invention of steam navigation, and when his fellow-citizens ascribed it to him he wrote to Paris and disclaimed it. To both of them alike all honour and gratitude are due.

The Académie des Sciences has recently, at the request of Mdlle. Marthe de Jouffroy, the grand-daughter of the illustrious inventor, appointed a committee to examine the question as to whether her grandfather is not entitled to some mark of national recognition; and this commission unanimously agreed to associate itself with the municipality of Besançon, in erecting a statue to one whose discovery was turned to material advantage by the foreigner, but which is none the less one of the glories of France.

CHAPTER IX.

ALGERIA AND TUNIS.

Si vis pacem, para bellum.

IN order to obtain the great advantages which the possession of Algeria insures to France, we must consider the difficulties or facilities which the character and habits of the Mussulman Arabs offer, regarded from the point of view of European civilisation.

I am not speaking of the results which must be attributed to Algeria in the military education of our army, of what relates to life in the open, the aptitude for enduring fatigue and privation, the value to our soldiers of struggles which, as in the Middle Ages, have an individual character. I am thinking more of the novel moral dispositions derived in Algeria from contact with the native populations.

In the early days of the conquest, the duty and the constant preoccupation of the French authorities were loyally to carry out the Convention of Algiers, which guaranteed to the Arabs that they should be allowed the free exercise of their religion, that their habits should be respected, and that they should be left in full enjoyment of their properties. The Arabs had struggled long and manfully against our rule, and it was to be feared that the war would leave feelings of rancour and prejudice in the breasts of those who might be appointed to administer the tribes after the pacification. But, by a happy selection, the army which had vanquished the natives was entrusted with the duty of governing them. It had learnt to appreciate what was honourable in their character; it had become initiated into their habits and language, and had opened its ranks to a large number of Mussulman soldiers. It was, therefore, in a position to fulfil the duty allotted to it not only with justice but, to its credit we may add, with generous sympathy for the vanquished.

Without being blind to the radical difference in feeling and aptitude which mark the two races, we have proved that there is no inseparable barrier between the Mussulman Arabs and ourselves, and that civilised Europe need not look upon them as incorrigible barbarians.

The Arabs who serve under our flag have gained a brilliant position side by side with our bravest troops. Under the conduct of the able officers who managed the Arab bureau, they built houses which they gradually began to inhabit; they planted trees, constructed dams, extended their areas of cultivation, improved their roads, and took the first steps towards the constitution of well-regulated civil life. When once we entrusted them with arms, the teaching and the example of the intrepid and kindly-disposed officers placed in command soon made excellent soldiers of them. When we shall have given them well-selected industrial leaders we shall derive immense benefits from the labour of these quick-witted Algerian races.

But in order to succeed it is indispensable to treat the Mahometans with the kindness and sympathy due to men whom we shall some day have to make French citizens. There has ceased to be any irreconcilable hatred between the Eastern and Western races; and it is for France to organise and administer with equity the Mussulmans subject to her authority. Fanaticism against the Christians no longer exists except among the Turks, for the Arab race, which follows the practices of Islam in all their purity, and according to the precepts of the Koran, regards as infidels the idolaters, and not the Christians.

France has governed Mussulmans for more than fifty years, and though many people regard them as subjects who are not upon equal terms with the French political family, I consider it as a civic duty not to withhold from them our solicitude and esteem. It would be very inconsistent for us to treat the Mahometans of Algeria as rayahs when we are urging the Sultan to emancipate the rayahs of the East.

We must not, in our relation with the Mahometans

of Algeria, lose sight of the real views of their apostle
in regard to the Christians—views expressed in the
Koran, though the meaning of them has been changed
by fanatic commentators. The proclamations which
Mahomet addressed to his compatriots, and which
have become chapters of the Koran, applied principally to the tribes of the Arabian peninsula, who
were given over to idolatry. He enjoined them to
respect the belief in the one God.

We read in chap. ii. verse 59: "Assuredly they
who believe and practise the Jewish religion, and
the Christians; in a word, all who believe in
God and do good works shall receive the reward
of the Lord; fear shall not fall upon them, and
they shall not be afflicted." Verse 25: "No con-
straint in matters of religion. The right path is
easily distinguished from the way of perdition."
Chap. iii. verse 78: "We believe in God, in what he
has sent us, in what he has revealed to Abraham,
Ismail, Jacob, and the twelve tribes; we believe in
the Holy Books which Moses, Jesus, and the prophets
received from heaven. We make no distinction
between them. We are resigned to the will of God."
Verse 98: "The Jews and the Christians believe in
God. They order all to do good and forbid that
which is evil. They vie in good works, and they are
virtuous. Chap. iv. verse 16: "But the men of
solid learning among the Jews and the Christians, as
well as the faithful, which believe in that which has

been revealed to thee and before thee, those who make
prayer and give alms, who believe in God and in the
day of judgment, to all them will we grant a glorious
reward." Chap. v. verse 7 : "This day you are permitted to do all that which is good ; you are permitted to espouse the virtuous daughters of the
faithful, and of them who have received the Scriptures
before, provided that you give them a dowry."
Verse 51 : "Let those who hold to the Gospel judge
according to its contents. Those who do not judge
according to a book of God shall be impious." Chap.
xxix. verse 45 : "Do not enter upon any controversy
with the men of the Scriptures, save in the most becoming manner, unless it be with the wicked. Say: We
believe in the books which have been sent us, as well
as in those which have been sent to you. Our God
and your God are one. We submit ourselves wholly
to his will." Chap. v. verse 35 : "He who shall kill
a man who has committed no murder or done no
wrong in a country, the same shall be regarded as the
murderer of the whole human race, and he who shall
have given back a man his life shall be regarded as
having given back the life of the whole human race."

It will be seen from these quotations that Mahomet
never anathematised the faith sanctioned by the
Pentateuch or the New Testament. He never spoke
of Moses or Jesus save in the terms of the utmost
veneration ; he never refused his benevolent protection to Christian priests and monks ; he never com-

manded intolerance or set an example of fanaticism. Before he began to preach, at the time when he was sent by his first wife, who was older than himself, to trade in Syria, he was the guest of the monks in the Holy Land, and he received the teaching, especially in matters of religion, from the monks who kept watch over the Holy Sepulchre. On returning to Arabia, he spent some time on Mount Sinai; and he was so grateful for the way he was treated during his twelve months' stay there, that he left with the Patriarch a document, at the foot of which he placed his hand dipped in ink by way of a signature. This document conveyed a grant to the Patriarch of Mount Sinai of certain privileges and of various properties in the region one day to be conquered by Islam. The grant was recognised as valid after the establishment of the Turks at Constantinople, and it is deposited in the Treasury at Stamboul. The concessions granted by Mahomet were carried out, and this was what made the Patriarchate of Sinai the wealthiest religious establishment in the East. Among the concessions granted by the Prophet was the produce of the customs at Suez. I discovered this little-known fact in the following manner. One day Said Pasha, the Viceroy of Egypt, who had granted me the concession, told me that he had purchased from the Patriarch of Sinai the Suez customs, which would, he added, be a profitable transaction if our enterprise succeeded.

Mahomet, in enjoining hostility against the infidels,

that is to say, against the idolators, had solely in view the pacification of Arabia.

In the seventh year of the Hegira, three years before his death, he meditated propagating the Islam faith beyond the frontiers of Arabia.

"The Mussulmans," says Rabasson, in his "Histoire de Charles Quint," "are the only enthusiasts who, by taking up arms to propagate the doctrine of their Prophet, have enabled those who refused to receive it to remain attached to the practices of their own worship."

When the Mahometans went to besiege Jerusalem, the Holy City offered a long and obstinate resistance. Finding at last that they could hold out no longer, the Christians agreed to capitulate, upon condition that they should treat with the Caliph in person. Omar, who had succeeded Abu-Bekr, the father-in-law and successor of the Prophet, having left Medina as soon as he was informed of this, proceeded to Djabia, where the Jerusalem delegates came to see him. He granted them the free exercise of their religion, and confirmed them in the possession of their churches. The Patriarch Sophronius received, upon entering Jerusalem, the chief of the Mussulmans, who, by the simplicity of his costume and the austerity of his life resembled more one of those Christian anchorites and dwellers in the desert than the prince of a people already famous for its victories. Omar went through several quarters of the city, with his hand linked in that of the Patriarch, and discoursing familiarly with him. The hour of prayer having come, he withdrew to the steps of the eastern portico of the church of Constantine, fearing that if he prayed inside the church the Mahometans would seize it and convert it into a mosque. Passing through Bethlehem, he prayed in the church built over the grotto where Jesus was born. But to prevent it being taken away from the Christians, he left a written order forbidding the Mussulmans to pray in it more than one at a time.

In Africa, the same spirit of moderation marked the progress of the Islam faith. When it made its appearance among the many heresies which were disgracing the African Church, it was regarded not so much as a new religion as a Christian sect. The partisans of Arius welcomed it almost, and it spread without persecution or violence among the barbarous tribes relegated to the southern countries after the recent invasions which had swept across Africa.

In Algeria, the Mussulmans must be treated as fellow-citizens, entitled to equal rights and equal respect, while in the East they must treat us as we treat their brethren in Algeria. What nonsense has been written about the intractable fanaticism of the Algerian Arabs! How often Abd-el-Kader has been represented as an implacable sectary! The people who made these accusations had never lived among the Mussulmans, or their acquaintance was limited to those who inhabited the towns, where the presence of the French had revolutionised all their habits of life, increased the friction, and engendered profound antipathy.

The opinion of those who have been in constant communication with the Arabs is, as a rule, very different. They have understood that fanaticism had not nearly so much to do with the resistance of the Arabs as patriotism. Religion was the only flag around which they could rally and concentrate their efforts, and it indisputably has been a powerful stimulant for inducing them to confront the perils of an unequal struggle, to support the evils of war, ruin, exile, and misery, though since December, 1847, when Abd-el-Kader declared it impossible to continue resistance, religion has not been for an instant an obstacle in the way of pacification. The exhausted tribes have accepted French rule ; the so-called fanaticism has disappeared, as if by enchantment, in the course of the relations which ensued on the establishment of peace; the taxes have been regularly paid; and the chiefs invested with authority have been universally obeyed.

This is not the place to explain the causes which have, on various occasions, interrupted these friendly dispositions, and led to severe repression, but something may surely be forgiven this grand people if they exhibit some little mistrust and irritability against

the conquerors of their country. After having combated them with the utmost energy, we cannot but
esteem them. Time, which heals so many wounds, is
speeding onward ; a sincere respect for their religion
and customs, great equity in our administration, and
a constant solicitude for the welfare of the people and
for their education, will aid us to conquer their hearts,
just as the bravery of our soldiers has overcome their
armed resistance.

I have mentioned the name of Abd-el-Kader. Those
who knew him during his captivity and in Syria,
where he saved the Christians from Turkish barbarity,
have admired the noble simplicity of his manners, the
even benevolence of his disposition, and the loftiness
of his mind and ideas.

He preserved his prestige undiminished, and whenever he came forward to express tolerant feelings in
the face of Europe, it was with the conviction that he
would not lose the confidence of his co-religionists.

A few years ago I wrote to ask him to send me a
circular, which had been addressed to all the Arab
chiefs of the region in which the late Commander
Roudaire was about to conduct his researches with
regard to the formation of an inland sea in the Tunisian
and Algerian chotts. His letters of recommendation
proved very useful, and facilitated the accomplishment of M. Roudaire's mission; and I trust that this
scheme, calculated to effect the pacification of Southern
Algeria and Tunis, will be carried out.

Subjoined are some extracts from an Arab work which Abd-el-Kader addressed a few years ago to the French Asiatic Society:—

"All the prophets, from Adam to Mahomet, are agreed upon the fundamental points : they have all proclaimed the unity of God, and the duty of paying him worship. . . . There is one point common to all—that of proclaiming respect for the divinity and charity towards His creatures. The modifications which have occurred, at different epochs, relate to principles of emergency, to matters which vary according to circumstances. Just as a doctor may prescribe one potion one day, and another the next, in the same way it may be said that a religion is good for the epoch in which it was revealed. Mahomet said, 'I am not come to abolish the Pentateuch or the Gospel, but to supplement them. The Pentateuch contains external directions appropriate for the masses; the Gospel contains inward directions specially intended for those who seek perfection. I admit both the one and the other; I maintain the *lex talionis*, which is a guarantee for the security of human life. So much for the external and general directions. At the same time I enjoin pardon for injuries received as an excellent means for being pleasing in the sight of God. So much for the inward and special precepts.'

"It will be seen that in reality these three religions are but one, and that the divergences between them are only on points of detail. One may compare them

to children of the same father by different mothers.
If the Mussulmans and Christians will be guided by
my advice, they will live in harmony and treat each
other as brethren, in speech as well as in outward
form."

The foregoing observations and quotations are made
by me with the view of contributing to the pacification of Algeria, which we hold by virtue of a conquest
which half a century's expenditure of blood and money
has legitimised.

With regard to Tunis, it is henceforward united
to France, under the sovereignty of the reigning
family, by the ties of a vassalage which dates,
morally speaking, from the conquest of Algeria, and,
materially, from the day when the Government of this
territory, which is wedged in, as it were, between our
possessions, endeavoured to shake itself free from our
preponderating influence.

I am one of the earliest participators in our constant
policy in this respect.   Going back to the capture of
Algeria in 1830, I will recall an incident not generally
known or remembered.   As soon as our troops had
taken possession of the provinces of Algeria and Oran,
the Government which succeeded that of Charles X.
declared in favour of a partial occupation.   It was
then that my father, Mathieu de Lesseps, Consul-
General and Chargé d'Affaires, with whom I was
serving as student-consul, bethought himself of asking the Bey of Tunis to authorise his brother and heir

to accept the Beylicate of Constantine, under the authority of France, and in consideration of a tribute guaranteed by Tunis. Taking with me this treaty, concluded *ad referendum*, signed by the Bey and the representative of France, I went with it to Marshal Clauzel, the Governor-General of Algeria, who approved its terms. Various circumstances prevented its ratification in Paris, but it none the less remained on record from this date that we could not under any circumstances allow the Bey of Tunis to place himself under the effective dominion of Turkey or any other Power, to the detriment of the security of our Algerian possessions.

CHAPTER X.

I HAVE spoken in the previous chapter of Abd-el-Kader, who for thirteen years maintained so gallant a struggle against the best of our African generals, until, hemmed in by superior force, he was compelled to surrender to General Lamoricière.

I will not attempt to describe his career in the field, but I am in a position to give some particulars as to the life he led after he had become our prisoner. When on my way to the Madrid Embassy in 1848, I stopped on the way at the Château de Pau, where Abd-el-Kader and the whole of his family were detained, I had never seen him before, and I was struck by his air of nobility and resignation. He spoke highly of the bravery and generosity of our army, and showed himself resolved to serve France as effectually by his moral influence as he had combated her bravely sword in hand. He was to his very last hour faithful to his promise. His conduct during the Syrian massacres in 1866 checked the excesses of the Mahometan fanaticism. Surrounded by his sons, he constituted himself the protector of the Christian population of Damascus, and his services were recognised with the ribbon of the Legion of Honour. In the year following, while travelling in Syria to get together labourers for the Suez Canal, I sent a message from Jerusalem to inform the Emir that I proposed to come and pay him a visit at Damascus, the inhabitants of which were reported to be still very hostile to Europeans. He came out to meet me as soon as my caravan was within sight, and made me mount beside him in his carriage. We then drove through the city, the inhabitants, who were drawn up in long lines outside their houses, prostrating themselves before him to the ground, and I spent several days there, being treated with great kindness.

It will be remembered that Abd-el-Kader came to Paris in the Exhibition year (1867), and was, with all the sovereigns of Europe, the guest of the Emperor. In 1869 he left Damascus to greet the Empress at Port Said, and to be present with her at the opening of the Suez Canal, when the French frigate *Forbin* was placed at his disposal.

Abd-el-Kader prolonged his stay in the Isthmus, where the Suez Canal Company gave him the use of the domain of Bir-abu-Ballah, at the entrance to the valley of Goshen, near Ismailia. One of our surveyors of works had built a pleasant house there, with gardens and land reclaimed from the desert. This territory formed part of the vast domain of Pithom, which had been purchased by the Canal Company, and upon the 25,000 acres of which ten thousand Arabs were already employed.

My intention was to obtain Abd-el-Kader's consent to superintend the cultivation of the 150,000 acres which had been conceded to us to the west of the Canal, from Lake Timsah to Suez, and through which we had already cut a sweet-water canal. But the policy which had in vain endeavoured to prevent the execution of the maritime canal still continued to stimulate the suspicions of the Viceroy of Egypt, who begged me to abandon my scheme, which I did with the concurrence of Abd-el-Kader, whose behaviour was, as usual, very loyal and disinterested.

When Commander Roudaire was charged by the French Government with the mission of completing his researches as to the possibility of making an in-land African sea, Abd-el-Kader, as I have already mentioned, sent a circular to the Arab chiefs, enjoining them to assist him. And when I recently under-took a voyage of discovery to the same region, the Emir sent me a fresh message, which may be regarded as a noble testament on his part, for he intended it to help to pacify our African possessions, and to attach to us by links of kindness the three million Mussul-mans who are subject to our laws.

" Praise to the only God !

" Abd-el-Kader ben Mahiddin, to all the Arab

tribes inhabiting Tunis, and more especially to their ulemas, sheiks, and religious and military chiefs.

"Salutation to you, with the mercy and blessing of God!

"While forming my wishes for your prosperity, and wishing you well in all that concerns the welfare of the body and soul, I take it as my bounden duty to give you the following counsel.

"The French Company which formed the project of piercing the Isthmus of Gabes, and concerning which I have already spoken to you, has now determined to put the work into execution, and to pay a visit to your neighbourhood. It is to be hoped, nay, it is your bounden duty, to see that these strangers meet with from you a most favourable welcome, generosity, encouragement, and assistance, both by word and deed.

"Do not lend the ear to those who erroneously imagine that the piercing of the Isthmus of Gabes is contrary to the interests of the country and of its inhabitants. These are false conjectures, and those who believe in them are ignorant people. Moreover, if it be God's will that this enterprise should be carried out it will be, however little it may be expected.

"It was thus that God permitted the piercing of the Isthmus of Suez, the benefits of which are now being reaped by humanity.

"In short, this French Company, the object of

which is to ameliorate the land by increasing its fertility and diminishing the extent of waste land, will do no harm to any one, and in the event of its requiring a field, a garden, or a house, it will pay for them a much higher price than they are worth. Moreover, Providence utilises this Company, which is by itself very powerful, thanks to the riches which God has granted it for the good and benefit of His creatures.

"It is for this that the Company is about to make great efforts, and spend immense sums in order to benefit the creatures of God. It is true that the Company will gain some fruit from its labours, but is it not also the creature of Allah?

" So it is with the king when he is just and good. Although he is the chief of his subjects, and placed in a position higher than they are (seeing that upon him depend the fertilisation of the country and the suppression of waste lands), he is in reality only the servitor of his subjects, and his duty is to seek to do them all the good he can, and guard them from all that is hurtful.

" For those who labour to this end a great reward is promised in heaven, but if they seek an earthly reward God will grant it to them here below; but if they seek a heavenly reward God will grant it to them in the other world.

"A prophet of the Israelites said, ' The kings of the Persians are heathens and fire-worshippers; they have been loaded by Thee with good things. They adore another than Thee, and yet Thou leavest them their kingdom and givest them long life.'

"And God said unto him, 'These people have made my land to prosper, so that my creatures can live therein with comfort. This is why I have left them their kingdom and granted them long life.'

"The prophet David built the holy temple at Jerusalem, but so it was that no sooner had he built the house than it fell to the ground. And God said unto him: 'Because thou hast shed blood abundantly, and hast made great wars, thou shalt not build an house unto my name.' *

"'But, O Lord!' replied David, 'is it not for Thy glory?'

"'Yes,' replied the Lord; 'but are they not my creatures whom thou hast slain?'

"Thus men are of the family of God, and the Lord loves those who seek to do good unto his family.

"The human race is very dear to God, its creator, and all His creatures, from the highest to the lowest, are meant for the service and benefit of the great whole which we call the human kind.

"ABD-EL-KADER EL HUSNY.

"The 23 Rébi-el-Anouar, 1300."

---

* See 1 Chron. xxii. 8.—Note of the Translator.

CHAPTER XI.

ABYSSINIA.

I.

Origin of the Abyssinian People.

THE Abyssinians have a tradition, the origin of
which is lost in antiquity and which is said also
to be prevalent among the Jews, viz., that soon after
the deluge, Chus, the grandson of Noah, went through
Lower Egypt, which was then uninhabited, and crossing the Atbara settled with his family in the tablelands of Abyssinia. The same tradition relates that
Chus and his family, still terrified by the recollection
of the Deluge, chose rather to live in caves upon the
mountain side than to trust themselves to the plains.

This race of men hewed with amazing perseverance
large caverns in the mountains of marble and granite,
many of which are still in existence.

The Abyssinians also say that the children of Chus
built the town of Axoum, shortly before the birth of
Abraham. Soon after this they established colonies
as far as the Atbara, where, as we gather from
Herodotus (Book II., chapter xxix.), they cultivated

the sciences. Josephus, in his "Antiquities of Judea," calls them Meroëtes, or inhabitants of Meroë (Atbara), an island situated between the Astaboras and the Nile.

The fragments of the colossal statues of the constellation of Sirius, which are still to be seen at Axoum, show that this people possessed some astronomical knowledge. Seir, in the language of the Chussites or Troglodytes and in that of the land of Meroë, means " dog," which explains why this province was named Siré and the large river which skirts it the Siris.

In the plain between the Fazoglou and Sennaar the river is named *Nile*, that is to say, *blue*. The ancients knew it by this name and also by that of Egyptus, but they more generally designated it by that of Siris. Pliny says that it bore this name above its junction with the other branch, that of the White Nile : " Sic quoque etiamnunc Siris, ut ante nominatus per aliquot millia et in Homero Egyptus."

The name of Egyptus, which Homer gives to the stream, was known in Ethiopia long before his time ; and Egypt in Ethiopian is called Y Gypt, while an Egyptian is Gypt. Y Gypt signifies the country of ditches or canals.

Thebes was built by a colony of Ethiopians who came from Siré, the city of Seir or of the dog-star, and of Meroë. Diodorus of Sicily says that the Greeks, by putting an *o* before Siris had made the word unintelligible. Siris then was Osiris, but he was

neither the sun nor a real person. It was the star
Sirius or the dog-star, designated by the figure of a
dog because of the information which it gave to the
people of Atbara, where were made the first observations of its emerging from the sun's rays which made
it easy of perception with the naked eye. The comparison of the " barking Anubis " was made because
its first appearance was like the barking of a dog
which gave notice of the approaching inundation.
The theory of the constellation of Sirius was specially
studied at Thebes on account of its connection with the
rural year of the Egyptians.

Ptolemy has related an heliacal ascension of Sirius
observed upon the fourth day of the summer solstice
in the year 2250 B.C.; and there are very good reasons
for believing that, long before this period, the Thebans
were excellent astronomers. This observation certainly makes Thebes much older than it is supposed to
be according to the chronicles of Axoum.

That city is not mentioned in the Bible by the
name under which it is known to us. Before Moses's
day it was destroyed by Salotes, Prince of the Agaazi
or Ethiopian pastors. In the ancient tongue it was
called Ammon-No. The name of Thebes is said to
be derived from Theba, a word which in Hebrew
signifies the ark (of polished wood, *theba*) which God
ordered Noah to build.

While the descendants of Chus were extending
their progress in the central and northern parts of

their territory, their brethren were advancing into
the mountains which run parallel with the Gulf of
Arabia. This country was always known as Saba,
or Azab, both of which words signify the south. It
was thus called because it was on the southern coast
of the Gulf of Arabia, and that, on coming from
Arabia or Egypt, it formed the southern frontier of
the African continent.

The inhabitants, who wore long hair and had very
delicate and regular features, with dark brown skin,
and who lived with their flocks in tents upon
the vast plains, made overtures to the Chussites
and acted as messengers to them for the conveyance of their merchandise. These men were called
Phut in Hebrew, or in all other languages, Balous,
Bagla, Belavé, Berberi, Barabra, Zilla, and Souah,
all of which signify pastor. The country which they
inhabited was called Barbaria by the Greeks and
Romans, after the word *Berber*, which originally
signified pastor.

It was over the long tongue of land which extends
along the shores of the Indian Ocean and the Red Sea
that the pastors carried the merchandise to the ports
of these two seas as far as the plains of the Isthmus
of Suez, which probably derives its name from Souah
pastors.

In the Bible one of these plains is spoken of as
Goshen, that is, the land of pasturage, and the Arabs
still call it Beled-el-Guéche, which means the same.

The principal residence of the pastors was the low and level part of Africa situated between the tropic of Cancer and the mountains of Abyssinia. But the noblest and most warlike of the pastors were, beyond all doubt, those who inhabited and still inhabit the mountains of Habad, which extend from the neighbourhood of Massowah to Suakim. In the ancient language of the country *so* means shepherd, *souah* being the plural.

The mountains inhabited by the Agaazi are called *Habad*, which in their language, as in Arabic, means serpent. Hence comes the historical tradition told in the book of Axoum, that a serpent conquered the province of Tigré and ruled over it.

According to this book, which is the most ancient chronicle in the country and the best authority next to the Bible, five thousand years elapsed between the creation of the world and the birth of Christ. Abyssinia was not inhabited until 1800 B.C., and four hundred years later many eminent men, speaking different languages, sought refuge there. They were well received by the Agaazi, and each one of them was allowed to choose the land which he wished to occupy. This establishment is called in the Chronicle of Axoum, *Angoba*, that is to say, the entry of the nations. There is a tradition, too, that this people came from Palestine at about the time that an inundation caused great damage there, and we know from Pausanias that there was a great inundation in Ethiopia during the reign of Cecrops in Greece in the year 1490 B.C. At this period the Israelites, leaving Arabia, entered the promised land under Caleb and Joshua. We cannot wonder at the terrible impression which this invasion made upon the minds of the dwellers in Palestine. Thus, when Joshua had crossed the Jordan and caused the walls of Jericho to fall, a panic seized all the peoples of Syria and Palestine. (See Joshua vi. 21.)

These peoples, each of whom spoke a different language, hearing that the conqueror, followed by a numerous army and already master of a portion of the country, was putting the vanquished to death beneath harrows of iron, did not wait to face so formidable a foe, and sought safety in flight, their most natural refuge being the pastors of Abyssinia and the Atbara. Procopius mentions two columns which in his day were still standing upon the coast of Mauritania, opposite Gibraltar, and upon which was inscribed in Phœnician, "We are Phœnicians, and we are flying before the face of the son of Nun " (Joshua).

Thus, among the various inhabitants of Abyssinia, from the southernmost limits to the frontiers of Egypt, there were to be found descendants of Chus, who, after having been troglodytes and lived in caves, and then pastors, became partially civilised and resided in cities. After them came the nations which left Palestine—the Amharas, the Agows of Damot, the Agows of Tohue, and the Gafats.

II.

Journey of the Queen of Sheba to visit Solomon at Jeru-
salem, and Conversion of Abyssinia to the Jewish
Faith.

It is not surprising that the constant traffic and the
important business transacted by the men of Tyre and
the Jews with the Chussites and the pastors of the
African coast should have established close relations
between them. We can understand, therefore, that
the Queen of Sheba, the sovereign of those lands,
should have desired to see for herself what became of
the treasures which had been exported in such large
quantities from her own country, and to make the
acquaintance of the prince for whom they were intended. There can be no doubt as to the journey
having taken place, for all the Eastern nations speak
of it in the same terms as those in which it is described in the Bible. The Abyssinian annals say that
the Queen lived at Saba or Azab, the land of myrrh
and incense, situated not far from the Red Sea. They
add that she went to Jerusalem under the auspices of
Hiram, King of Tyre, whose daughter accompanied her,
as we are told in Psalm xlv. ; that she did not go by
sea or pass through Arabia for fear of the Ishmaelites,
but went from Azab into Palestine, and returned by
way of Massowah and Suakim, escorted by her own
subjects, the pastors ; and that she performed the

journey upon a white camel or dromedary of very great size and surpassing beauty.

Many ancient writers imagined this queen to be of Arabic descent; but Sheba was a kingdom of itself, and must not be confounded with a small town in Arabia also called Saba, to the south of Mecca. We know from history that the Sabeans were accustomed to be governed by a queen rather than a king, while the Homerites, or Arabian Sabeans, who inhabited the coast of Arabia opposite Azab, were ruled by kings. The Homerite kings were not allowed to leave their country or even their residence, and if they appeared in public the people had a right to stone them.

We may be sure that a people which treated its sovereigns in this way would not have allowed the queen, if perchance they were ruled by one, to undertake a long journey. The Arabs assert that the name of the Queen of Sheba who came to Jerusalem was Belkis, while the Abyssinians call her Maqueda. In the New Testament, Matthew speaks of her as the Queen of the South (chap. xii. v. 42).

The annals of Abyssinia are full of details concerning her journey. They say that the queen, who was a pagan when she left Azab, was so filled with admiration of Solomon, that she became converted to the Jewish faith while at Jerusalem, and had a son by Solomon, whom she named Menilek. The queen brought him back with her to Sheba, but a few years afterwards sent him to his father to be educated.

Solomon was careful to give him a very good education, and he was anointed king of Ethiopia in the Temple, taking henceforth the name of Solomon's father David (Daoud). He then returned to Azab with a colony of Jews, among them many doctors of the Mosaic law, including one of each tribe. He made these doctors judges in his kingdom, and from them are said to be descended the present judges (umbares), three of whom always accompany the king. With Menilek was Azarias, son of the high priest Sadoc, bearing a copy of the law; and he, too, was given the title of Nebrit, or high priest, while, although the book of the law was burnt in the church at Axoum, when the Arabs despoiled the province of Adel, the functions of Azarias were preserved in his family, his descendants being still nebrits, or priests, of the church of Axoum.

The whole of Abyssinia was thus converted to the Jewish faith, and the government of the state as well as of the church was modelled upon that of Jerusalem. The last use which the Queen of Sheba made of her power was to order that no woman should in future reign, and that the crown should go to the nearest heir male. In the later history of Abyssinia we find that if no woman wore the crown, many queen-regents have left a great name behind them, and it may even be said that the most prosperous and peaceful epochs of Abyssinian history have been when a queen was regent. The Queen of Sheba died after a reign of forty years, about 986 B.C., and was succeeded by her son Menilek, whose descendants were, as we know from the traveller Bruce, still on tho throne in 1790.

III.

Conversion of Abyssinia to Christianity.

The Abyssinians accept the Holy Scriptures as we do, and count the same number of books.

The Revelation of St. John, called by them "the Vision of John-Abu-Kalamsis," is their favourite reading. The old Abyssinian priests read with much gusto the Song of Solomon, but they prohibit the reading of it to their deacons, to laymen, and to women. They believe that Solomon composed it in honour of tho daughter of Pharaoh. Next to the Revelation they esteem the Acts of the Apostles, which they style *Synnodos*, these *Synnodos* serving as the written laws of the country.

Another book is called *Haimanut-Abu*, and consists chiefly of the works of Greek fathers treating of and expounding certain articles of faith which were the subject of disputation in the ancient Greek Church. There are also translations of the works of St. Athanasius, St. Basilius, St. John Chrysostom, and St. Cyril also extant in Abyssinia. Another book much revered is the *Synaxar*, or "Flower of the Saints."

According to Abyssinian history, Bazen, who was the twenty-second king descended from the Queen of

s 2

Sheba, was contemporary with Augustus and reigned sixteen years, the birth of Christ taking place in the eighth year of his reign. The cónversion of Abyssinia to Christianity took place under King Abreha, or Atzeba, the thirteenth successor of Bazen, about 333 years after Christ, and the first Bishop of Abyssinia was delegated by St. Athanasius of Alexandria, who himself occupied the episcopal see. of that city, A.D. 330.

It is also related that Frumentius, the apostle of Abyssinia, came to the kingdom during the government of a woman, who was probably the mother of a king under age. The Greek philosopher Meropius, who was living at Tyre and had embraced the Christian religion, embarked upon the Red Sea to go to India, taking with him Frumentius and Adesius, two young men whom he was anxious to establish in trade, after having given them the best of educations. The ship upon which they had embarked was wrecked off the coast of Abyssinia, and while Meropius perished in defending himself from the inhabitants, the two youths were captured and taken to Axoum, where the court then resided. They soon became acquainted with the language, and as the Abyssinians were always very kindly disposed towards strangers, they were very well treated, Adesius being appointed master of the king's household, a post which has since then always been held by a foreigner. Frumentius was deemed worthy to be entrusted with the education of the king, and the queen appointed him her son's tutor. Frumentius inculcated in him great veneration and love for the Christian religion, and he then proceeded to Alexandria to inform Bishop Athanasius of his hope of converting Abyssinia to Christianity, and to ask him to send there a number of men capable of spreading instruction among the people.

Athanasius consecrated him Bishop of Axoum, and on his return the king publicly embraced Christianity. The greater part of Abyssinia followed his example, and the Church of Ethiopia has endured down to our own day.

It appears that the conversion took place peaceably and without any effusion of blood. This was the second time that the empire changed its faith in the same orderly fashion, no fanatical preachers or over-zealous saints causing any disturbance. If war has at various periods desolated Abyssinia, it has been for purely temporal reasons.

Towards the year 1200, while Lalibala reigned in Abyssinia, the Christians were violently persecuted in Egypt. Amru, the lieutenant of the Caliph Omar, had then completed the conquest of that kingdom, and the masons and stone-cutters suffered more than any of the others, as the Arabs had a special detestation of those trades. Lalibala offered many of them a refuge, and employed them in hewing out of the solid rock in the province of Lasta, his native place, a number of churches which are still intact.

During the reign of Saif-Araad (of the line of Solomon), from 1342 to 1370, the Soudan of Egypt had imprisoned Mark, the Patriarch of the Copts, and as soon as Saif-Araad heard of it, he ordered all the Egyptian merchants to be arrested, and sent bodies of cavalry beyond the frontier to stop the caravans. The Soudan soon released the Patriarch, the only condition he stipulated being that he should make peace between him and the Abyssinian king, which he soon did.

Zara Jacob, fourth son of David II., succeeded his nephew and occupied the throne for thirty-four years (1434—68) under the name of Constantine, and he was regarded in Abyssinia as a second Solomon. The Abyssinians had a long time before this founded at Jerusalem a monastery, to which Zara Jacob made several donations, and he obtained permission from the Pope to found a second one at Rome. Nicodemus, then superior of the monastery at Jerusalem, sent priests in his name to the Council at Florence, and these priests concurred in the views of the Eastern Church as to the procession of the Holy Ghost, which was the cause of the schism between the Greeks and the Latins. The Abyssinian embassy was deemed of sufficient importance for the recollection of its visit to have been preserved in a picture which is still in the Vatican.

IV.

*Struggle of Abyssinia against the invasion of the Mussul-
man tribes of Arabia and the coast of Africa. Its
alliances with Portugal. Before and after the discovery of the Cape of Good Hope.*

Prince Henry, son of John I., King of Portugal,
jealous of the greatness of Venice, which owed its
prosperity to the trade with India, discovered another
means of communicating with the East, and that was
by sailing round the famous cape then known as the
Promontory of Tempests.

He had to combat the prejudices of the whole
nation, but he had learned from history that the
voyage had already been accomplished by the Phœnicians, during the reign of Necos in Egypt, and
afterwards by Eudoxius under Ptolemæus Lathyrus.
Eudoxius passed round the southernmost point of
Africa and arrived at Cadiz.

But there are always plenty of people who, incapable of achieving any great thing themselves, are
ready to criticise the enterprise of others, and these
people declared that the sea was continually raging
and boiling around these arid shores, and that the air
was so heated by the sun that all men who went
through it would come out quite black. These arguments, industriously circulated by the Venetians,
would have sufficed to prevent Prince Henry's project
being carried out if King Edward, instead of being

influenced by them, had not favoured his uncle's plans, and several voyages were made under his auspices.

Christians returning from Palestine reported that they had seen in Jerusalem a monastery, the monks in which were subjects of a Christian prince in the heart of Africa, whose empire extended from the shores of the Red Sea and the Indian Ocean to the shores of the Atlantic. It was further said that several of these monks came to Alexandria, the patriarch of which alone enjoyed the privilege of sending a bishop into their country. This Christian prince was known in Europe as Prester John. While sending vessels to circumnavigate Africa, the King of Portugal despatched two ambassadors to Prester John by way of Egypt. Covillan and Païva were entrusted with this mission, and they took with them a map drawn by Prince Henry, being instructed to correct it by the light of the observations which they made.

The Portuguese travellers went together to Alexandria, Cairo, Suez, and Aden, where they separated. Covillan proceeded to Calicut and Goa, and from that point, crossing the Indian Ocean, visited the mines of Sofala. On his return to Aden and Cairo, where he was to have been rejoined by Païva, he learned that the latter had died.

At Cairo he received the visit of two Jews, Abraham and Joseph, bringing two letters from the King of Abyssinia, into whose states he then made his entrance. King Alexander received him with great kindness and kept him at his court. The ambassador married an Abyssinian woman, and was in high favour with several of the princes who succeeded one another upon the throne. He kept up a correspondence with the King of Portugal, describing to him the different parts of India which he had seen, the wealth of the Sofala mines, to the north of the Cape of Good Hope, and exhorted him, on his own behalf as well as that of the King of Abyssinia, to persevere in his researches as to the feasibility of a passage round the Cape. He assured him that the possibility of it was well known in India and Abyssinia, and sent him a map upon which the Cape and the country round were correctly drawn.

Thereupon, the King of Portugal fitted out three vessels which he placed under the command of Bartholomew Diaz, who reached the formidable cape, but his sailors, terrified by the force of the wind and the rough seas, refused to go any farther. The sailors, whose complexions were burnt brown by the sun and the long sea voyage, were afraid of becoming literally Negroes. All the stories which had been told them before their departure appeared to them as realities, and Diaz was obliged to content himself with seeing the Cape of Good Hope, instead of sailing round it, returning to Portugal, where, for the remainder of the king's life, the dangers of the expedition were being constantly dwelt upon.

In order to divert the king from carrying out his spirited enterprise, many influential persons, including the envoys of foreign sovereigns, based their opposition upon motives of state policy. They urged, as it has since been urged in regard to the Suez Canal, that the enterprise was an impossible one, and that as, in the event of its succeeding, the balance of trade would be altered, the nations which had the exclusive possession of the trade with India would combine in a war of extermination against Portugal.

Prince Henry was no longer alive to answer these contradictory objections and perfidious suggestions, and since then the spirit of enterprise and maritime discovery had declined in Portugal.

But some years later King Emanuel determined to follow up the noble project of his predecessors. He selected as his lieutenant Vasco de Gama, a man of great distinction both as regarded his courage and general disposition, and he intrusted him with the journal and maps of Pedro Covillan, as well as the letters of the African and Indian princes of whom he had heard.

Upon July 14, 1497, Gama started from Lisbon with a small fleet, and upon the 18th of November he discovered the Cape of Storms. But the ships were so tempest-tossed that the sailors refused to go any farther. The impressions made by the voyage of Diaz were stronger than the obedience and resignation which they had solemnly sworn in the Chapel of the

Virgin, to which Vasco de Gama had conducted them
in procession before he left Lisbon. They revolted,
the pilots placing themselves at the head of the
mutineers. But Vasco, seconded by his officers, seized
the leaders of the revolt, and loading them with irons,
placed them in the hold. He himself went to the
helm, and, steering off the land, went out to sea, to
the great astonishment of his brave companions. The
tempest lasted two days more, and on the 20th of
November he had the honour of being able to say that
he had doubled the Cape. In the moment of victory
the trumpets were sounded, and Vasco liberated the
prisoners, amid great rejoicing, and impressed upon
them that the proper name for the promontory was
the Cape of Good Hope.

The admiral landed with Martin Alonzo, who spoke
several of the Negro dialects, upon the Tierra de Natal,
where he was very well received by the king and the
natives.

Upon the 15th of January, 1498, after having taken
in a fresh supply of water, which the Negroes themselves helped him to get on board, Gama proceeded as
far as a cape which he named the Cape of Currents,
where the coast of Natal commences, that of Sofala
being farther north. He reached the very spot
where Covillan, coming from the north, had previously arrived, so that these two Portuguese went
right round Africa.

David III., the ancestor of Alexander, ascended

the throne in 1508, when twelve years of age, the Queen-Regent, Helena, and Bishop Mark, her favourite, assuming the reins of government in Abyssinia, which began to suffer from the attacks of the Mussulman kings upon the eastern coast of Africa and on the Arabian side.

Helena, the daughter of a Moorish prince, did all she could to keep the peace between the Abyssinian Christians and their Mahometan neighbours by creating business relations between them, and she had succeeded to a great extent, when a third Power came in to disturb the equilibrium. The Turks, who had never appeared in the south of Africa or Asia, came upon the scene, under Selim, the Emperor of Constantinople, who had just conquered the Soudan of Egypt, soon establishing themselves in the Arabian peninsula up to the shores of the Indian Ocean.

The leading towns on the coast of Arabia—Jeddah, Moka, Suakim, and Massowah, upon the African coast, at the gates of Abyssinia—were garrisoned with Turkish janissaries, who preyed upon commerce instead of protecting it, so the Arab traders took to flight, going with their riches to the coasts of the kingdom of Adel, upon the south-eastern limits of Abyssinia. The trade of India, in order to avoid a like hindrance, was also concentrated upon Adel.

The Turks then seized Zeyla, a small island situated upon the coast of Adel, at the entrance to the Indian Ocean, where they established a custom-house and subjected the trade of the kingdom of Adel with India to heavy dues. This new establishment threatened both the kingdom of Adel and the empire of Abyssinia, and the Queen-Regent Helena, hearing of the Portuguese passage round the Cape, saw that nothing but their assistance could save Adel and Abyssinia from ruin. Pedro Covillan, the Portuguese, was still at her court, and she arranged with him to form an alliance with the King of Portugal. There was also at her court an Armenian merchant named Matteo, who had a great reputation for probity, and who had been in the habit of travelling through the Eastern States to fulfil missions for the kings and the great. Helena selected him as her ambassador to the King of Portugal, and it appears certain that the despatches which he carried were drawn up by Pedro Covillan, their contents being that the Queen's demands would be explained in person by Matteo, who enjoyed her full confidence.

Ambassadors travelled more slowly in the sixteenth century than they do now, and Matteo first went to the Portuguese Indies, it being only three years afterwards, in 1513, that he continued his voyage to Portugal, whither he proceeded with a fleet loaded with spices sent home by Albuquerque, the Portuguese Governor-General.

During this time Helena had concluded a treaty of peace with the King of Adel, but as the relief expected from Portugal did not arrive, that prince, incapable of resisting the Turks, allied himself with them against

Abyssinia. Their combined forces invaded the empire, and in less than a year they had reduced to captivity or had slaughtered twenty thousand Christians. The whole country was terrorised, but David III., though only sixteen, placed himself at the head of an army, while the Queen-Regent and the ladies of the nobility freely contributed their jewels and were lavish in presents to the soldiers, in order to stimulate their courage. The King soon reached the province of Fategar and marched direct upon Aoussa, the capital of the kingdom of Adel. There he drew up his army in battle array, and after a single combat between a young Abyssinian monk, Gabriel Andreas, and Maffudi, one of the Adel leaders, in which the latter was killed, a great battle was fought, in which the Abyssinians were victorious, ten or twelve thousand Moors being left on the battle-field. The next day King David went to a city where the King of Adel had a palace, and finding the gate shut he struck it with his lance. No answer being given, he left his lance sticking in the door, to show that he had come hither and had been free to enter the gates. When the army returned to Abyssinia the young monk who had been the hero of the single combat was loaded with honours, his victory being commemorated in songs. This victory was gained on July, 1516, and upon the same day a Portuguese fleet, under the command of Don Lopez Suarez Alberguiera, had seized the island of Zeyla and burnt the custom-house. The ambassador Matteo, who had

been right royally treated by King Emanuel and sent
back to the Indies, embarked at Goa with Admiral de
Segueyra and sailed for Massowah, where he arrived
on April 16th, 1520. He then set out for the interior
of Abyssinia; but the fatigues of the voyage had been
too much for him, and he died of fever before he could
regain King David. Zaga Zaab, an Abyssinian monk,
was selected as his successor, and he started for Portugal in 1525, the year of the death of Queen Helena.

David then made preparations for renewing the war
with the kingdom of Adel, which had allied itself with
the Turkish pashas and generals commanding in Arabia,
the Turks sending a contingent which began by re-
capturing the island of Zeyla.

It was customary for a caravan to go every year
from Abyssinia to Jerusalem, this caravan—which consisted of about a thousand pilgrims, priests as well as
laymen—starting from Hamozem, a small territory only
two days' march from Dobarwa and Massowah. The
caravan was preceded by trumpeters, and crossed the
Desert by way of Suakim without meeting with any
rebuff. But in the year following the conquest of Egypt
by the Sultan Selim, when the reign of the Mameluke
dynasty ended the Abbot Azerata-Christos was conduct-
ing fifteen hundred pilgrims to Jerusalem, and on their
return, having been met by a body of Selim's troops,
most of them were massacred and the rest driven into
the Desert, where they perished of hunger and thirst.
In 1525 another caravan assembled at Hamozem. It

was composed of three hundred and thirty-six monks
or priests and fifteen nuns. The second day after it had
started it was attacked by the Moors of the Hamozem
district, and all the Christians of a certain age were put
to the sword, the younger ones being sold to the Turks.
Only fifteen persons escaped, of whom three alone succeeded in rejoining the king at Shoa.   From this time
the Abyssinians cut off all communication with Egypt
by way of the Desert, and David entered with his
army the province of Dawaro, sending on a detach-
ment of troops which defeated the Adel advance guard,
while the king advanced and fought a great battle at
Chimbra-Coré, in which he was totally defeated, losing
a great part of his nobility and four thousand soldiers.

Mehemet, surnamed Gragne (the left-handed), Governor of Zeyla, was in command of the allied army,
and he spent the next two years following this victory
in strengthening his forces, at the expiration of which
time he invaded the frontier provinces of Fategar,
Efat, and the Dawaro, putting most of the inhabitants
to the sword and reducing the remainder to slavery.

Seeing his empire threatened with ruin, King
David resolved, despite his inferior forces, to fight
another battle, but he was once more defeated, losing
his principal commander and leading officers. He
returned to Amhara and encamped at Hegis, hoping
to recruit a fresh army, but the Turkish commander
did not give him time to do this, and in the month of
April following entered Amhara and then burnt and

pillaged Varvar. In 1530 he invaded the province of Tigré and the King fled to Wogora, while in the year 1531 the Abyssinian king, still pursued by Mehemet, sustained a third defeat at Dalakas, on the banks of the Nile.

Negadé-Yasus and many other heads of the nobility perished beneath his eyes, and the brave monk, Andreas, now well advanced in years, sought a glorious death, being resolved not to survive the disasters of his country.

Other disasters followed, but King David continued an heroic resistance until his death in 1540, the final blow to him being the capture by the Turkish Vizier Mudjid of the whole of his family, who were put to the sword.

The only one who escaped was his son Claudius, and when he came to the throne the fall of the Abyssinian empire seemed inevitable, especially as famine and pestilence, which generally followed upon a prolonged war in the East, were desolating the country. Claudius, who had been very carefully brought up by his mother, Sabel-Venghel, celebrated for her wisdom and courage, had not, of course, much experience, and the Moors thought that they would soon have Abyssinia at their mercy; but the young sovereign had the good fortune to beat them in several engagements, and finally overthrew the army of Ammer, the principal lieutenant of Mehemet.

While this favourable change was taking place in

Abyssinia, the Patriarch Juan Bermudez, who had been sent several years previously to negotiate an alliance with Portugal, returned from Lisbon, and he drew so graphic a picture of the disasters of Abyssinia that the king sent orders to the Viceroy of the Indies to send four hundred soldiers to Massowah. Don Stephano de Gama, brother of Vasco, who was Viceroy of the Indies, determined to land Juan Bermudez with the promised troops on the coast of Abyssinia, and his fleet passed through the Straits of Bab-el-Mandeb to Massowah. These troops, commanded by Martin Correa, seized the town of Ashiko and put all the inhabitants to the sword, Martin Correa cutting off the head of the Moorish commander and sending it as a present to Queen Sabel-Venghel, who was at that time residing in a fortress of the kingdom of Tigré.

Don Stephano de Gama, returning to India, left his younger brother, Christopher, behind him with some of the best of the Portuguese troops, and the latter, after combating the Turks with varying success, was eventually made prisoner by the treachery of a Mahometan woman with whom he had fallen in love, and delivered up to Mehemet the Left-handed, who had his head cut off and sent it to Constantinople, his body being divided among the tribes of Arabia.

Mehemet also seized the Portuguese camp and allowed his men to despatch all the wounded, but when the Turks pursued the women to the lines of

Don Christopher, where they had sought refuge, one of them, to avoid the outrages to which they were about to be subjected, set a light to a powder barrel and blew up the whole camp. The Queen and the Patriarch succeeded in making their escape, and rejoined King Claudius, who was very grieved when he heard of Christopher's death. He soon avenged it, however, defeating Mehemet the Left-handed in a battle at Bet-d'-Isaac, on February 10, 1543. Mehemet himself was killed by a bullet fired by Pedro Leon, a Portuguese, who cut off his ear and put it in his pocket, returning to the ranks to continue the fight. The Moors, deprived of their general, took to flight, and were pursued until nightfall by the Abyssinians and Portuguese, who slaughtered them in great numbers.

Thus Claudius took a splendid revenge upon the Mussulmans who had reduced his father to such cruel extremities, and it only remained for him to punish Joram, who had driven his father from Mount Salim and compelled him to cross the Tacazzé on foot at the risk of being drowned. Joram was not at the battle of Bet-d'-Isaac, but he hastened to march in that direction, and the king, informed of his intention, put some of his troops into an ambuscade and cut Joram's army to pieces.

While Mehemet had been ravaging Abyssinia, the provinces of Siré and Tigré, situated between the Demba and the cities which the Moors occupied on the Red Sea, had been the theatre of the war. The Turks had completely ruined them, and Mehemet had burnt the city of Axoum and destroyed all the churches and convents of Tigré, Claudius being occupied during the end of his reign in repairing these disasters.

But Del-Tumborea, the widow of Mehemet, did her best to keep up the war, for she told Nur, the Governor of Zeyla, who was madly in love with her, that she would only give her hand to the man who brought her the head of Claudius, the conqueror of Mehemet.

Nur eagerly accepted the challenge, and sent a message to Claudius, bidding him defiance. Claudius quickly reassembled his army and marched upon Adel, contrary to the advice of the queen-mother and his friends, who advised him to wait the coming of the Moors. The battle was a very bloody one, but the Abyssinians were worsted, and Claudius succumbed after receiving twenty wounds. His head was cut off and brought by Nur to Del-Tumborea, who had it suspended by the hair from a tree facing her house, in order that her eyes might ever be able to feed upon a spectacle so grateful to them.

Claudius had reigned nineteen years, and the battle in which he perished was fought on March 22nd, 1559. The principal officers of his army perished with him, and a great part of the army was made captive, the remainder being dispersed and the camp pillaged. Nur, content with the recompense of his undertaking, did not care to renew the struggle, and he returned to Adel attired as a private soldier, forbidding any of the demonstrations which usually greet a victorious soldier, and declaring that the glory of the triumph was due to God alone.

Since that time the Moors have scarcely ever interfered with the Abyssinian empire, and the reigns of the kings of the Solomon dynasty who succeeded Claudius, from 1559 to 1770, were marked by a series of rebellions, of internal struggles, and of wars, many of them unsuccessful, with the Gallas tribes bordering on Abyssinia.

v.

Modern and Contemporary Period.

At the end of the eighteenth century the governors of the principal provinces refused obedience to the monarch descended from Solomon. The princes of that family had lost their authority, and, up to the present time, Abyssinia has been governed by the ras or kings of the two large divisions which form the empire of Abyssinia: Tigré and the Ambara.

Tigré, with its dependencies, comprises all the region between the Red Sea and the Tacazzé. The Ambara, with its dependent provinces, is formed by the territories between the Tacazzé and the Nile. In 1855 an Abyssinian chief, who was merely governor of a province, who did not belong to the Solomon race, revolted against his father-in-law, Ras Ali, who had been reigning for a long time at Gondar. He overthrew him, and after having vanquished first Oubié, King of Tigré, and then the King of Shoa, proclaimed himself Emperor under the title of Theodoros. But, as we have seen was the case in previous ages, Abyssinia, a mountainous country favourable for defence as well as for attack, has been the scene of many sudden changes in the fortune of war. In 1858 and 1859, Theodoros was in his turn defeated by Negoucié-Nikar, a nephew of Oubié, who regained possession of forty-four provinces forming part of the kingdom of his uncle, while his brother Dedjammadjé-Tassamma, took possession of Gondar, the second city of the ancient empire. A relative of Ras-Ali, named Amadin-Bechir, several times defeated the army of Theodoros, and remained in possession of the provinces of Wollo, Warro-Cassou, and Warro-Imanat; the King of Shoa recovered his independence by forming an alliance with Amadin-Bechir and another chief, named Tedela-Gualu, who governs the provinces of Godjam, Damot and Agos-Meder, up to the sources of the Blue Nile, while the Gallas tribes are constituted into a kingdom and are hostile to Theodoros.

Thus, having regard to the number and importance of the provinces which he has reconquered, King Nikas seems to be the most powerful prince in Abyssinia, and I trust that this unfortunate country, which has been subjected to all the horrors of civil war, may recover the unity which in former ages saved it from foreign conquest, and that King Nikas may, by his intelligence and tendency to open communication with Europe, be equal to this difficult but glorious task.

He sent me, I may add, the following autograph letter, upon his own behalf and that of his people, expressing his wishes for the success of the Suez Canal; and this letter, written in Ethiopian, has been translated by M. d'Abbadie, well known for his travels in Abyssinia.

"I Negus,

"Master (of the horse) Nikas, King of Ethiopia, who reigns by the law of our Lord Jesus Christ, from Mizwa to Gondar, and this is the kingdom of Tigré, and Simen, Wagara, Walqayt, Tagadé, Dambya, Balasa, Kinfaz, Agaw Lasta, Salawa; I salute Ferdinand de Lesseps, who is of the tribe of light, who has accomplished a work wonderful for our day.

"From the beginning until now I have had my mind fixed upon the work which you are accomplishing, and which is a source of joy for all the earth; and now that it is a settled thing, upon behalf of my country, which I love, and in my own name, I give you thanks.

"In piercing the land of Sawis (Suez), you make a mutual union between our lands and the affairs of

Europe. Thus your name will not perish from among us, and our land will be a granary for the regions of the West. And this being so, know that my country and I love you. I am anxious to aid you in your enterprise with cattle or in any other way. I pray the Lord to keep you."

I complete this plain narrative of the leading facts in the history of Abyssinia, by expressing the hope that France will come to an understanding with England to restore to a population of thirty million Christians, now driven into the mountains, their ancient maritime territory.

France has respect for all forms of religion, but she is opposed to religious fanaticism, and it seems to me that in what looks like the impending disturbance of the Mahometan world, she has a noble mission to fulfil, that of maintaining aloft the standard of civilization in the vast regions of Algeria, Senegal the Gaboon, the Congo, Christian Ethiopia, and the Roudaire Sea.

CHAPTER XII.

THE inhabitants of Marseilles and the Catalonians were the first commercial people in Europe who, after creating consuls, at first merely the syndics of the principal trading corporations, and afterwards judges in matters of local trade, felt the importance of extending the influence of this institution abroad. The "consuls beyond the seas" were thenceforward entrusted with the duty of keeping a watch upon the privileges of their nation, and of settling all disputes between fellow-countrymen in regard to matters of trade. Their duties were considered very important, and were entrusted to men who apparently belonged to the leading families in the county.

It was during the Crusades that French princes entrusted to the maritime towns and nations which assisted them, principally to the inhabitants of Marseilles and the Catalonians, the privilege of forming in the conquered ports corporations of traders, under the control of the consuls of their nation. The first privileges obtained in Syria by the inhabitants of Marseilles date from 1117—1136.

The Marquis of Montferrat, Seigneur of Tyre, gave permission in 1187 to the Marseilles traders in that city to appoint a consul to dispense justice.* Three years after this, Guy de Lusignan allowed this city of Marseilles, by letters patent, to appoint at Acre consuls or viscounts, who were sworn in by the King of Jerusalem, and who had jurisdiction in all civil and criminal cases, murder and high treason excepted.

Although at this period, Marseilles had no foreign consuls in the city, her own magistrates took special care of the interests of foreign traders. In her municipal statutes (*statuta civitatis Massiliæ*) drawn up in 1228, 1233, and 1255, Marseilles laid down as a principle that, even when at war with a city or a State, it was the duty of the adversary to respect the private property of the inhabitants of that city or State—a principle which does honour to the city which proclaimed it. Avignon, following the example of Marseilles, had also declared the property of strangers to be inviolable, in time of war as well as of peace.

In 1148, the town of Narbonne possessed at Tortosa, in Spain, a commercial establishment, and the privilege of having a consul there; while similar privileges had been obtained by Narbonne at Genoa in 1166 and at Pisa in 1171.

* See *Histoire du Commerce entre le Levant et l'Europe,* by Depping.

A traveller in the fourteenth century found at Alexandria a French consul whose mission it was to protect the foreigners who had no consul of their own nationality.* This honourable privilege of protecting the foreigners who had no consul of their own has been confirmed by the treaties styled *capitulations*, concluded between France and the Ottoman Porte, as far back as the reign of François I., treaties by which the protection of the Catholics is accorded to France.

Jacques Cur took advantage of his position at the court of Charles VII. to give a sort of official character to the relations which he had for some time established in Egypt. The Sultan, flattered by his presents, wrote in 1447 a letter the king, in which he promised his protection to French traders, and authorised the appointment of a consul, whom he agreed to treat upon the footing of the most favoured nation.†

Barcelona, the neighbour and rival of Marseilles, soon entered into competition with her for European trade. Gradually delivered from the yoke of the Sarrazins, from the end of the ninth century, by the assistance of France, she commenced, under Raimond Beranger, towards the close of the eleventh century an era of great prosperity. Her maritime trade had then acquired sufficient importance to elicit the en-

---

* Extract from Fuscobuldi, quoted by Pardessus in his *Introduction aux Lois Maritimes.*

† *Mémoires de Mathieu de Coussi*, quoted by Pardessus.

couragement of the Sovereign, who had the wisdom to guarantee protection and assistance to all foreign ships, even to those of the Sarrazins. The thirteenth century is the epoch in which the Catalonian trade made its greatest advance. The relations between the Catalonians and France were very important; they attended the fairs in Champagne, and, as we learn from Pardessus's "Collection des Lois Maritimes," they maintained a consul there. Thus the capital of Catalonia, which has provided maritime and commercial legislation with the celebrated "Consulate of the Sea," showed as keen an appreciation as Marseilles of the usefulness of foreign consulates.

James I., King of Aragon, granted in 1266 to the municipal magistrates of Barcelona the privilege of annually electing and sending out to Egypt and Syria consuls of their own; and towards the close of the fourteenth century the Catalonians drew up some regulations for the consulate at Alexandria, according to which the consul was appointed for three years and was re-eligible. He was forbidden to keep a tavern or sell wine by retail, to let the shops on the ground-floor to any but Catalonians, or to admit into his house Jews, or women of ill-fame. He was to be present all day at the custom house, if required, to take part in the examination of goods, and whenever he left his house he was to be preceded by two men in livery.* As early as the thirteenth century, the

* See Capmany's *Mémoires Historiques*, vol. xi.

Catalonians had consuls at Constantinople, Beyrout, Damascus, Cyprus, Rhodes, &c., and they had one upon the confines of Asia, at Tanaïs, who in 1397 appeared before Tamerlane and offered him presents upon his return from the triumphant expedition into Muscovy and Kipsac.*

In Europe, the Catalonians had consuls among all the peoples living upon the Mediterranean : at Marseilles, Genoa, Pisa, Naples, Venice, and Sardinia, and especially Sicily. They also had a consulate at Seville, and the historian Capmany mentions fifty-five consulates of which Barcelona could boast in the days of her splendour, but of which not more than five or six remained in the sixteenth century.

In the Act of Privilege which King Ferdinand granted in 1251 to the Genoese at Seville, it was especially stipulated that the Genoese should have in that city consuls of their own nationality, with the right of deciding without appeal all disputes between persons of their own nationality. If the dispute was between a burgher of Seville and a domiciled Genoese, it was also to be settled by the consuls, but an appeal was to lie to the alcaldes. The consuls were to have nothing to do with criminal affairs.†

The habit of appointing consuls in a foreign

---

* See Count de Laborde's *Itinéraire d'Espagne*, vol. v.

† See Navarreté's *Coleccion de los viages y descubrimientos que hicieron por mar los Españoles*, vol. xi.

country did not become general until the sixteenth
century, and especially from the reign of Louis XIV.,
and in the course of a short time all the trading
nations sent consuls to one another and conferred
upon them prerogatives more or less extensive.

Colbert was the true organiser of the consulates,
and his memoir of March 15, 1669, "upon the steps
to be taken by consuls of the French nation abroad
to keep his Majesty informed of all that occurs," was
the first outcome of the measures which this enlightened Minister had adopted for improving the
consular institution.   Soon afterwards, the fundamental ordinance of 1681, which was also his handiwork, placed the consulates in a position to render
genuine service to French commerce, and formed for
more than a century the legislation by which French
consular establishments were governed : up to the
reforms which were commenced in 1803 and have
been gone on with ever since.

The Spanish Government has not yet carried out
its project of publishing a set of rules in which the
ancient ordinances relating to consulates will be
fused, in order to provide a general body of instructions for all its agents.

The general purpose of a consul is to act as commercial agent for his Government in a foreign port or
place of trade, to keep an eye upon the commercial
interests of his country, to endeavour to develop them,
and above all to uphold before the local authorities

the rights of his fellow-countrymen and to arrange their disputes.

There are two kinds of consuls, one being delegated by his Government to exercise a special jurisdiction over his compatriots and their business affairs, without having any other character than that of magistrate and public functionary, while the other is a trader who is allowed to add to his particular profession the duties of consul.

There are several reasons for preferring that a consul should have no interest of his own in commerce. His time and his labour should be not his own, but should belong to his country and Government, to which, like the traveller Anacharsis, he should communicate all that it may be desirable to know concerning the laws, the customs, the habits, the arts, the trade, and the manufactures of the country in which he lives.*

According to the general instructions for French consuls in foreign countries signed by Louis XVIII. in 1814, " the consuls are political agents, but only in this sense, that they are recognised by the Sovereign who receives them as officers of the Government which sends them, and that the principle of their mandate is either specific treaties, or the common custom of nations or general public law."

Then, again, the preamble of the ordinance of December 15, 1815, says : " Consulates being insti-

* See the Comte de Gardens's, *Traité de Diplomatie*, vol. i.

tuted to protect the trade and navigation of our subjects in foreign jurisdiction, to exercise justice and control over our said subjects, and to supply the Government with information which may enable it to insure the prosperity of foreign trade, we have recognised the fact that this object cannot be attained if the persons selected for the duties of consul have not acquired by special studies adapted for the character of their work, as well as by a certain amount of experience, a thorough knowledge of public law, of legislation, and of commercial affairs."

This rule, though at times disregarded, was confirmed by the royal decree of August 20, 1833, relating to the personal composition of the consulates, the fifth clause of this decree providing that "Consuls-General are to be selected from among the first-class consuls, the latter from among the second-class, and the latter from among the students for consulships," the only exception being in favour of the clerks employed in the commercial branch of the Foreign Office after so many years' service.

That learned jurisconsult, M. Pardessus, in his "Cours de Droit Commercial" (Part VII. chap. vi.), has devoted several chapters to the political character of consuls, their jurisdiction, the various administrative or mixed functions which are conferred upon them, and to the punitive rights which in certain cases they have as against individuals of their own nationality. There is nothing, however, to prevent a Government conferring upon its. agents such an amount of latitude as may be deemed compatible with its interests, and they should be considered as public officials if they devote their attention solely to public affairs, and if the Sovereign who appoints them, and whose subjects they are, confers this rank upon them.

It is only in the Levant and in Barbary that the consuls have a right of absolute jurisdiction over their compatriots. In other countries they must confine themselves to jurisdiction in trade disputes, which is usually conferred upon them by treaty and usage; and if they decide as to the personal differences between their compatriots, it can only be when called in to arbitrate.

Of all the conventions concluded between the European Powers, none better defines the rights, the immunities, the privileges, and the duties of consuls than that concluded between France and Spain on March 13, 1769. This convention and the previous treaties between France and Spain, as set forth in ministerial decrees and royal ordinances, empowers consuls "To collect and administer the property of their compatriots who have died *ab intestat*. To exercise the full authority conferred by the navigation laws over the vessels of their own nation. To regulate the salvage of shipwrecked vessels. To claim the surrender of deserters from ships. To assist at the examination of trading vessels where their intervention, or that of one of their agents, is deemed indispensable. To be present when the houses of any of their compatriots are searched for contraband goods (their presence must be first requested by the local authorities before they proceed to the search). To act as interpreters for their fellow-countrymen, and to settle their differences by arbitration and not otherwise, and jurisdiction is specifically denied them by the treaties and by the tenor of their letters of *exequatur*. To appoint vice-consuls for the different ports in their districts."

The convention of 1769 also accords to the respective consuls, when they are subjects of the prince who appoints them and when they are not in trade : "1st. Personal immunity from being arrested or cast into prison, except for some atrocious crime. 2nd. exemption from all personal charges or service, and from having soldiers billeted on them. 3rd. The inviolability of their papers and those of their chancelleries, which are not to be touched under any pretext whatever, unless the consul is a merchant. 4th. The privilege of not being liable to be called as witnesses in court, the tribunal of war, or, failing it, the ordinary tribunal, in the event of requiring any judicial declaration from the consul, being expected to send him a polite message to say that they are under the necessity of coming to his domicile for that purpose, etc. 5th. The right of placing upon the door of their house a tablet representing a ship, with the inscription, 'Consul of France or Spain.' "

In a speech delivered on the 3rd of March, 1838, at the Académie des Sciences Morales et Politiques, Prince de Talleyrand, who had to pronounce the eulogium of Count Reinhard, who had been consul, director of foreign affairs, and ambassador, said, "How many things a man must know to make a good consul, for his duties are endless in their variety, and quite of a different character from those of other officials of the Foreign Office ; they demand a mass of practical knowledge for which special education is required. Consuls should be able to fulfil, in the event of necessity, the duties of judge, arbitrator, and reconciler. They must be able to do the work of a notary, sometimes that of a commissioner of the navy. They have to look after sanitary matters, and from them is expected, owing to their general relations, a clear idea of the state of trade and navigation, and of the industry peculiar to their place of residence. Thus, M. Reinhard, who took the utmost care to be accurate in the information which he was able to give his Government, and in the steps which he had to take as consular and political agent, and as administrator of the navy, had made a profound study of general and maritime law. This study had led him to the conclusion that a time would come when, by carefully prepared combinations, a general system of trade and navigation would be established, by means of which the interests of all

nations would be safeguarded, and the bases of which
would be such that war itself would not affect its
principle, even if it had the effect of temporarily
suspending its application."

The representatives of the Powers to whom the
Vienna Congress of 1815 attributed a diplomatic character, are ambassadors, ministers plenipotentiary,
resident ministers, chargés d'affaires. Since then,
this recognition has been extended to consuls-general
in those residences where they enjoy the title of agent,
as for instance, in Egypt.

Prince Metternich, in the course of a conversation
which I had with him at Vienna, towards the close of
his life, expressed views similar to those of Talleyrand,
adding, "Politics are a science, diplomacy is an art."

Prince Talleyrand, who was unquestionably an incomparable diplomatic artist, said in his speech at the
Institute:—

"Diplomacy is not a science of ruse and duplicity.
If straightforwardness is of prime value anywhere, it
is in political transactions, for it is that which renders
them solid and durable. People have confused reserve
and ruse. Straightforwardness is incompatible with
ruse, but it is not inconsistent with reserve, which,
indeed, strengthens the feeling of confidence."

M. de Talleyrand died three months after making
these remarks at a sitting which excited considerable
interest. The *Moniteur Universel* gives the names of
the principal members of the Institute who were pre-

sent, viz., Royer-Collard, Quartremere de Quincy, Bassano, Guizot, Thiers, Mignet, Cousin, Villemain, Lemercier, Molé, Fauriel, Montalivet, Sainte-Aulaire, de Barante, de Jaucourt, de Flahaut, Bertin de Vaux, de Noailles, de Valencay, etc. M. de Talleyrand entered the room leaning on the arm of M. Mignet.

XIII.

M. DE LESSEPS, having been elected by the French Academy to the chair left vacant by the death of Henri Martin, took his seat for the first time on the 23rd of April, 1885, and delivered the following address :—

"In admitting me among you, you have both conferred upon me a great satisfaction and placed me in a position of great embarrassment. To form part of the French Academy, this distinguished assembly, this elective aristocracy of letters, is an honour of which the proudest is entitled to be proud; but to speak before it is a task which may make even a clever writer hesitate, and I am, unfortunately, neither the one nor the other. The reception speech was therefore doubly formidable, both for me and for you. This is why I am anxious at the outset to reassure you. You are not about to hear a piece of oratory. I would not subject either my inexperience or your forbearance to so rude an ordeal. Unable to do well, I have done better—I have studied brevity.

"Your ancestors had the habit of summoning to the Academy of Letters, not merely men of letters, but men of mark—prelates, generals, and great nobles, whose high position was a substitute for eloquence, and sometimes even for knowledge. Was this why the speech delivered in such cases was reduced to the narrow proportions of returning thanks? Possibly. In any event, seeing that you have revived for me the first part of this tradition, allow me to benefit by the second; and seeing that you have been good enough to let a man of tetters *in partibus* enter, as formerly, your society, do not be surprised if he confines himself, as formerly, to a simple expression of gratitude.

"The chair I now occupy is the one successively occupied by M. Thiers and M. Henri Martin. Both were friends of mine—which is tantamount to saying that I am not ignorant of the dissimilarity between them and me, or of the distance which separates us. They were chiefly men of study; I am chiefly a man of action. They were historians, and I am a geographer—after a fashion. But if I differ from them on many points, there is one on which I claim to resemble them. Both passionately loved their country, and in that respect at least I do not feel myself unworthy to succeed them. Like them, I have devoted my entire life to my country. For more than sixty years, in various situations and with various fortune, anxiety for its interests and glory has been my ruling idea, the constant aim of my labours, and finally, as I am confident, the cause of my success.

"And such an aim required long exertions. Nothing is easy in this world, especially the useful. There is no fresh task, however beneficent, which has not, perhaps in the very ratio of the good it may do, the ignorant and the malevolent for enemies; the former, because they fail to understand the result which you are aiming at, and are not in the secret of your means or strength. They have to be enlightened. Once converted, they become fervent disciples and valuable auxiliaries. As to the others, the sceptics, the haters, even the insulters, they deserve no attention. The Arab proverb says, 'The dogs bark, the caravan passes.' I passed on.

"If I thus explain my views to you with an emphasis which may seem complacent, it is not for the empty pleasure of talking of myself; it is to justify you in your own eyes for having chosen me by showing the similitudes existing between my predecessor and myself. And as I am on this point there is one more which I wish to notice in passing. Both of us were accused, at starting, of a little too much imagination. You know that in the poetical and ardent moments of youth, and when entering on the study of the early times of our race, Henri Martin—so at least it is said—was smitten with the Druidical religion. This Celt of St. Quentin had been initiated, it is alleged, into the mysteries of the terrible religion. He was even suspected of having secretly embraced it and of practising its rites in private. Is this true or false? Did

he go through this excess of enthusiasm and convic·
tion ? It is far from certain ; but what matters it ?
In any case, did it prevent his writing later on the
most complete history of France yet issued ?

" As for me, if I am not suspected of being a Druid,
I was formerly charged with being a dreamer. This
was the beginning of my enterprises. I fancy that I
have since proved myself a practical man. I do not,
however, for all that, disparage dreamers. A little
imagination is a good leaven for the heavy dough of
human affairs. The more distant the goal the higher
you must aim. It is well for the sculptor to seek a
mountain for cutting out his first statue. It is not
amiss for the positive man to have to throw off his
mind a little of the impracticable and unattainable.
It is not amiss that, fancying himself omnipotent, he
has thought of attempting everything. Experience
will only too soon cut off what was impossible in his
illusion ; but his works will always retain something
strong and forcible to support them, his intelligence
something grand to elevate it. From the St. Simonian
aberration, now happily forgotten, there nevertheless
sprang accomplished engineers, distinguished economists, and first-class financiers. M. Thiers began by
writing criticisms of pictures. Claude Bernard himself, your illustrious colleague, began with a tragedy.
You have not that, at least, to twit me with.

"I spoke just now of the history of France, written
by Henri Martin. It is his chief work ; it is in all

the libraries; better still, it is in the memories of all. I do not wish to dwell on its literary merits, not because there is any lack of good to be said of it, but because I fear I should not say it sufficiently well. Besides, it is not a speech that I am now making, and if I stop to mark with a word what seems to me the special note of his talent, it is because it is at the same time that of his character.

"Each historian has his peculiarity. M. Michelet has poetry. Every moment his imagination opens wide views over new horizons, through which the mind roams in amazement. Augustin Thierry, an enthusiastic scholar, of a race of writers who called back to life a world that had passed away, is above all things a painter in clear lines, with an incomparable gift of colour. The history of Guizot, like that of Mignet, is a system, philosophical in the one, political in the other, showing in the movement of the facts their sequence, their consequences, and their causes. Thiers excels in recounting events, in bringing situations clearly before us, in elucidating the most special and most obscure questions. His ruling quality is clearness. That of Henri Martin is justice.

"And this love of justice which is in his mind comes from the love of country which is in his heart. Although a man with convictions, even a party man, absolute in his faith, invariable in his conduct, he puts aside all passion when he enters into history. A sympathising witness of all our glories, he withholds

his admiration from none. He is as enthusiastic an
admirer of the Druids as of the martyrs of the first
Christian Church, of Jeanne d'Arc as of Henri IV., of
the victories of Louis XIV. as of those of the First
Republic, of the First Empire as of the Convention.
For him it is France that is concerned, and her only
he sees. No restriction checks his patriotism, no calculation diminishes it. Whatever be their opinions
or their beliefs, all those who serve and benefit France
are his friends. This is a fine example to recommend
and to follow. Woe to those peoples who, driven to
fanaticism by party spirit, mutilate their traditions,
not understanding that a nation is a being, never
ceasing to live, whose present cannot be separated
from its past without existence itself being arrested.

"And from this past, so mournful and so glorious,
Henri Martin draws an invigorating lesson; a confidence that nothing will repress, a hope that nothing
will discourage : 'The Frenchman who knows the
history of his country,' he says, ' will never lose hope
in the saddest days. This people is endowed with an
incomparable spring of life, with a power of renovation which has never been met with to the same
degree in any other people.'

"I am proud that you have thought of me to succeed to the man who uttered such words. This double
sentiment of pride in the past and of faith in the future
is as deeply rooted in my heart as it was in his. It is
by this community of hope that I am proudest of re-

sembling him. And having now sufficiently proved my good will, in default of talent, I stop, not wishing to exceed the limits which I was, in a measure, bound to lay down for myself. He who will succeed me, following the more recent practice, will speak to you hereafter — as far hence, I assure you, as I can make it—with more fulness, competence, and charm of the merits of the impartial historian, the honest man, the great patriot, who was my predecessor. He will doubtless express to the Academy better than I can his gratitude as a newly elected member; but he will not have, at the bottom of his heart, more respect than I have for the memory of Henri Martin, or more gratitude towards you. In 1834, on entering the Academy, M. Thiers said, ' I thank you for having admitted me to a seat in this asylum of free and quiet thought.' I thank you, in my turn, for having admitted me into this asylum of free and quiet thought, although I do not promise to remain quietly seated in my chair."

Reply of M. Renan.

M. Renan, the Director of the Academy, spoke as follows in reply:—

"Monsieur,—Your address is charming, for it is your very self. I may tell you that we were not quite easy in our minds while you were preparing it, being afraid lest, for once in your life, you should deem it incumbent upon you to make a literary composition.

Your exquisite tact has saved you from making this mistake, and I detect in the tone of your observations the geniality and the contagious warmth which are the charm of your conversation. I was sorry to note the absence of certain anecdotes which are familiar to you, and I miss, for instance, certain details that you know about Abraham and Sarah, and about Joseph and the Queen of Sheba. A number of things which you know more about than any one else are absent from your speech, but nothing which is yourself is missing. You possess the greatest and the rarest literary quality of the present day—that of being natural. You never went in for declamation. Your eloquence consists of that manly and straightforward way of communicating with the public of which the example has been set by England and America. No one, assuredly, in our age, has been more persuasive than you, and in consequence no one has been more eloquent. Yet no one has taken less account of the artifices of language, or of the empty forms which are animated by no ardent conviction.

"You remarked upon one occasion: 'I approve of Latin and Greek being taught to our children, but what we must not neglect is to teach them to think wisely and to speak bravely.' That is what I so admire. You abhor rhetoric, and you are perfectly right. Rhetoric is, in addition to poetics, the only error we have to reproach the Greeks with. After having produced masterpieces themselves, they thought that they could lay down rules for others to do the same, in which they were much mistaken. For there is no more an art of writing than there is an art of speaking. To speak well is equivalent to thinking aloud. Success either in speaking or writing has never but one cause —absolute sincerity. When you excite the enthusiasm of a meeting and succeed in reducing that which is the most obdurate thing in the world to metaphors, and the most refractory to the artifices of the so-called art of fine talking—I mean capital—it is not your words but your individuality which attracts; or, I should rather say, your whole person speaks; you exercise a charm. You have that supreme gift which works miracles, like faith, and which is in truth of the same order. Charm has its secret motives, but not its definite reasons. Its action is wholly spiritual. You obtain the same amount of success at Chicago, a city which is not a third your age, as you do in the ancient cities of Europe. You convince the Turk, the Arab, the Abyssinian, the Paris speculator, and the Liverpool merchant, by reasons which differ only in appearance. The true reason of your ascendancy is that people detect in you a heart full of sympathy for all that is human; a genuine passion for ameliorating the lot of your fellow-creatures. You have in you that ' Misereor super turbas ' (I have pity upon the masses) which is the sentiment of all great organisers. People love you and like to see you, and before you have opened your mouth you are cheered. Your adversaries call

this your cleverness; we call it your magic. Ordinary
minds do not understand the seduction exercised by
great minds. The fascination of the magician escapes
the vulgar mind; the qualities of enchantment are a
gracious gift, and because they are imponderable
mediocrity denies that they exist, whereas it is the
imponderable which does in reality exist. Humanity
will always be led by the secret love-philters of which
the crowd sees only the superficial effects, just as the
illuminant of the physical world is in the invisible
fluids which the ordinary eye cannot discern.

"Your eloquence has captivated the whole world,
and has surely entitled you to a place in our midst.
The programme of our company is not a purely literary
one, carried out with no ulterior aim, and ending in
the frivolities which proved the ruin of Oriental litera-
tures. It is things or deeds which are beautiful;
words in themselves have no beauty outside the noble
or true cause which they serve. What matters it
whether Tyrteus was a man of talent or not. He
succeeded, he was as good as an army. The *Marsel-
laise* is, whatever musicians and purists may say, the
greatest song of modern times, inasmuch as it leads
men on to combat and to victory. When we reach
these altitudes personal merit is of small account; all
depends on predestination, or on our success, if that
word be preferred. It is no use saying that a general
ought to have won a battle if he loses it. The great
general—and this applies equally to politics—is the

man who succeeds, not the one who ought to have succeeded.

"Thus, the persons who were at first surprised to hear of your election were but very imperfectly acquainted with the spirit which governs our company. You have cultivated the most difficult of styles—one which has for a long time been abandoned among us— that of action. You are one of the small band of those who have maintained the ancient French tradition of a brilliant and glorious existence, one useful to all your fellow men. Politics and warfare are too lofty applications of the human intelligence for the Academy ever to have passed them over. Marshal de Villars, Marshal de Belle-Isle, Marshal de Richelieu, and Marshal de Beauvau, had no more literary titles to election than you have. They had won victories. Failing this qualification, which has become a rare one, we have chosen the master *par excellence* in the art of overcoming difficulties, the hardy speculator who has always won his wager in the pursuit of the probable; the virtuoso who has practised with such consummate tact the great and lost art of life. If Christopher Columbus lived in our day we should have made him a member of the Academy. The man who is quite certain to be a member is the general who one day brings back victory to our standard. We shall not quarrel with him as to the nature of his prose, and shall regard him as a very fit member of the Academy. We shall elect him by acclamation, without concern-

ing ourselves as to his writings. What a splendid
gathering that will be when he is received! In what
demand seats will be, and lucky the academician who
presides on that occasion !

" You have been one of those fortunate workers who
seem to have been taken into the confidence of what
the genius of civilisation requires at a given moment.
The first duty which man has had to impose upon himself in order to become in reality master of the planet
which he inhabits has been to rectify, in view of his
requirements, the combinations, in many cases opposed
to these requirements, which the revolutions of the
globe, ignoring altogether the interests of humanity,
have inevitably produced. What would have been the
fate of our planet if the parts of it which emerge had
been much smaller than they are ? if the field of evo-
lution of terrestrial life had not been larger than Easter
Island or Tahiti ? What historical fact has ever produced such consequences as that action of the sea
which suddenly brought Cape Gris-Nez and the cliffs
of Dover into being, and created France and England
by separating them ? Sometimes beneficial, these
blind chances of unforeseeing nature are sometimes
also very baleful, and then it is the duty of man, by
skilful readjustment, to rectify the evil services which
the blind forces of ancient times have done him. It
has been said, and with much truth, that if physical
astronomy possessed sufficiently powerful means, we
should be able to judge as to the more or less

advanced civilisation of the inhabited worlds by the criterion as to whether their isthmuses were pierced or not.

" For a planet is only ripe for progress when all its inhabited parts have reached that stage of close relationship which constitutes a living organism, so that no one part can enjoy, suffer, or act without the other parts feeling in harmony. We have reached that critical stage in the history of our own planet. Formerly, China, Japan, India, and America might have been convulsed by revolution without Europe so much as knowing of it. For long centuries the Atlantic divided the habitable globe into two parts as distinct one from the other as if they were two different worlds. Now, the stock exchanges of Paris and London are affected by what occurs at Pekin, in the Congo, in Kordofan, or in California ; there are but few dead parts in the body of humanity. The electric telegraph and the telephone have annihilated distance as regards the things of the mind, while railways and steam navigation have multiplied tenfold the facilities of bodily movement. It was inevitable, therefore, that our century should regard as an essential part of its work the removal of the obstacles to rapid communication. It was impossible surely that the generation which had tunnelled the Mont Cenis and the St. Gothard should be arrested by a few sandbanks or reefs of rock at Suez, Corinth, and Panama !

" You, sir, have been the chosen artisan for this

great work. The Isthmus of Suez had long since been selected as that the piercing of which was the most urgent. Antiquity had pointed it out, and had attempted the enterprise, but with insufficient means. Leibnitz had indicated this work to Louis XIV. as one worthy of his might. But the completion of such a work demanded a faith which the seventeenth century did not possess. It was the French Revolution which, reviving the age of fabulous expeditions and a state of heroic youth in which man is guided in his adventures by the flight of birds and the signs in the heavens, propounded this problem in such a shape that it could no longer be left dormant. The piercing of the isthmus was part of the programme which the Directory set before the Egyptian expedition. As in the time of Alexander, the conquest of arms was also the conquest of science. Upon December 24th, 1798, our illustrious colleague, General Bonaparte, started from Cairo accompanied by Berthier, Monge, Berthollet, and other members of the Institute, as well as by merchants who had obtained leave to follow in his escort. On the 30th he lighted, to the north of Suez, upon the vestiges of the old canal, and he followed them for a distance of more than twelve miles. On January 3rd, 1799, he saw, near Belbeys, the other end of the canal of the Pharaohs. The researches of the Egyptian Commission have formed the basis of all subsequent investigations, and the only point in which they were defective was the view as to the difference

in level of the two seas, a theory always opposed by
Laplace and Fourrier.

"The great St. Simonian School, which had so lofty
a comprehension of the common labour of humanity,
took up the idea and sealed it by martyrdom.   More
than twelve engineers of the St. Simon School died of
the plague in 1833, at the great dam of the Nile.
Amid much that was obscure and visionary, one great
truth was perceived, viz., that Egypt has an exceptional place in the history of the world.   The key to
the interior of Africa by the Nile, it is by its isthmus
also the guardian.   The most important point in the
empire of the seas, Egypt is not a nation, it is
a State, sometimes the recompense of a maritime
dominion legitimately won, sometimes the punishment of an ambition which has not measured its
strength.

"A country which has such important relations to
the rest of the world cannot belong to itself.   It is
neutralised for the benefit of humanity.   The national
principle is put to death there.   We are surprised to see
among the mad thoughts which passed through the
mind of Nero, during the hours which separated his
fall and his death, the idea of going before the people
in a suit of mourning and asking them, in exchange
for the Empire of Rome, to give him the Prefecture
of Egypt.   It is a fact that the Prefecture of Egypt
will always be an independent one.   The ruler of
Egypt will never bear the same name as other Sove-

reigns. Egypt will always be governed by the
civilised nations collectively. The reasonable and
scientific explorers of the world's history will always
turn with curious, anxious, or attentive glances towards this wonderful valley.

"France for three-quarters of a century has had
before her mind a solution of this difficult problem,
which will be fully appreciated when experience has
shown what torrents of blood and tears the other
solutions would have cost the world. She conceived
the idea of establishing, by means of a dynasty,
Mussulman in name, but in reality free from fanaticism and prompt to recognise the superiority of the
West, the reign of modern ideas in this exceptional
land, which cannot, without great detriment to the
general welfare, be allowed to lapse into barbarism.
Through Egypt, thus organised and safeguarded,
civilisation had her hand upon the whole of the
Eastern Soudan. The dangerous cyclones which
Central Africa will from time to time produce, since it
has imprudently been allowed to become Mahometan,
would have been suppressed. European science had a
free hand in a country which has, so to speak, been
placed at its disposal as a field for study and experiment. But there should have been something like
method observed in carrying out this excellent plan.
We should not have weakened a dynasty by means of
which the point of the sword of Europe reached
almost to the Equator. More especially should an

eye have been kept upon the Mosque of El-Azhar, the centre from which the Mussulman propaganda has been spread all over Africa. Isolated and abandoned to fetishism, the races of the Soudan are of little account; but converted to Islam they become *foci* of intense fanaticism. From want of foresight, we have allowed an Arabia much more dangerous than the true Arabia to be formed west of the Nile. Has it not surprised you, Sir, that there is not yet in that spot a common *sensorium* of the great interests of the world? It is clear that there is a guardian angel of humanity who prevents it from stumbling into all the ditches lining its way. If there were only diplomatists, I would as lief see our poor species intrusted to the care of a band of truant schoolboys.

The origin of your enterprise dates from the commencement of that dynasty of Mehemet-Ali which saw the light under the auspices of France, and which, upon the other hand, has been severely shaken by a passing declension in the fortunes of our country. Your father was the first French agent who resided in Egypt after the departure of our army. He was charged by the First Consul and by M. de Talleyrand with the task of counterbalancing the tyranny of the Mamelukes, which had the approval of the English. Your father's chief of the Janissaries brought to him one day, as being capable of combating the prevailing anarchy, a young Macedonian who was then in command of a thousand Albanians, and upon which the French expedition had made a very deep impression. This compatriot of Alexander could neither read nor write. His fortune grew rapidly, and as he forgot nothing which had been done either for or against him, when you arrived in Egypt at the beginning of 1832 as a student-consul, the powerful Viceroy at once took you into favour. Mohammed Saïd, one of his sons, was your early friend. You took a strange hold over him, and when he came to the throne you reigned conjointly with him. Through you he dimly perceived an ideal of light and justice for which his soul thirsted, but which dark clouds, issuing from a deep abyss of barbarism, still veiled for a time from his eyes.

You have described, in that easy and natural style which is all your own, the details of this intimacy which has been so big with consequences of the gravest import to the whole world; you have told us of his strange alternations of passion and good sense, of the enthusiasm for science of a nature but just removed from absolute ignorance, of the torrents of tears which succeeded his outbursts of mad fury, of the peals of laughter and of his ungovernable vanity : in short, of the struggle between a Tamerlane and a Marcus Aurelius. Your account of the wonderful journey which you made with him to the Soudan is a document of incomparable value for the student of Oriental psychology. The story of how, upon one occasion, he threw his sword across the room, fearing that in a moment of passion he might strike you with it, and how, when he had calmed down, you found him in tears because you had anticipated him in suggesting ideas of reform, is typical of the Eastern despot. The barbarian is always more or less of a child, and Mohammed's friendship was a glass which the least pressure of jealousy might break. You felt this, and your well-stored and supple mind provided for every contingency. It is only men of strong character who can deal properly with barbarians. Saïd had taken with him a service of Sèvres china for his own use, and he had given you another for your own use. The Viceroy's service, for want of proper care, was soon broken, while yours was intact. This would never do, so upon one occasion the well-trained camel which bore your service was replaced by a very skittish and almost savage camel. You were too sensible to remonstrate, and in a few minutes your service of china was broken to bits. The Viceroy nearly cracked his sides with laughter, and the work of the isthmus was safe. For from this period the piercing of the Isthmus of Suez was your constant preoccupation, and you had almost succeeded in getting your all-powerful friend to embrace your idea. Your views in this matter dated from an incident which followed your arrival in Egypt. You came from a country which had a clean bill of health to a country infested with disease; and, in obedience to a rule of logic which has never been altered, you were compelled to submit to a long quarantine at Alexandria. M. Mimaut, the French consul, to beguile the tedium of your confinement, brought you the great work published by the Egyptian Commission, specially commending to your notice Lepère's treatise upon the junction of the two seas. It was in this way that you became acquainted with the isthmus and its history. Henceforward the ambition to realise what others had conceived took hold upon you, and though you had to wait twenty-three years, nothing rebuffed you. You were born to pierce isthmuses, and antiquity would have made a myth of you. You are the man of our age upon whose forehead is most clearly written the sign of an unmistakable vocation. The principle of great deeds is to take possession of force where it is to be found, to purchase it at its proper price, and to know how to make use of it. In the present condition of the world barbarism is still an immense depôt of living forces. Your keen and open intelligence saw that immense power is often invested in hands incapable of making use of it, and that this power is at the disposal of any-one who knows how to employ it. You frankly take human affairs as they are. You do not mind the contact of stupidity and folly. It is all very well, you say, for those who do not˙ touch the realities of life to be fastidious and to remain immaculate. Humanity is composed of two thousand millions of poor ignorant creatures, to whom a small band of the elect, marked with a sign, are to impart reason, justice, and glory.

"Avaunt with the faint-hearted and the fastidious! avaunt with the over-nice, who would fain emerge without a speck of mud from the battle with stupidity and evil! They are not fitted for a work which demands piety rather than disgust, a proud and lofty heart, true kindness, which often differs very much from superficial philanthropy; something, in short, of the wide-embracing sentiment of Scipio Africanus, who, in reply to some trivial cavilling said: 'On such and such a day I won the battle of Zama; let us go up to the Capitol and return thanks to the gods.'

"It is to the East that you owe that gait, as of the Arab horse, which has sometimes startled your more timid friends. The East inspires a craving for grand adventures, for in the East the era of grand and fruitful adventures is not yet run out. The sight of sheep without a shepherd inspires one with the idea of taking charge of the flock. How often in Syria I have envied the sub-lieutenant who accompanied me! It may be that the man who is destined to found order and civilisation in the East is even now growing to manhood in some cadet school. You avoid in your appreciation of man the narrow judgments of implacable idea-mongers, who believe that all races of men are of equal value, and of flint-hearted theorists who see no necessity for the humble in the scheme of creation. Those people of the Lake of Mensaleh, who constructed the banks of your canal by gathering up the mud in their large hands and squeezing the water out of it against their chests, will have their place in the kingdom of God. Inferior, no doubt, they are, these poor human families, so cruelly treated by fate, but they are not, on that account, excluded from the common work. They may produce great men, and sometimes with one sudden bound they outshine us; they are capable of prodigies of abnegation and devotion. Such as they are, you love them. You are an optimist, Sir, and you are quite right. The height of art is to work good with evil, to achieve what is great with mean materials. This transcendant game is to be won by the sympathy and the love which one feels for men and which one inspires in them for oneself, by the audacity with which one persuades oneself that the cause of progress is gained and that one is contributing to it. The men of the East are above all things susceptible to being charmed, and you succeeded admirably in this. Your frankness and ease of demeanour inspired them with unbounded confidence. Saïd could not live without you. Your perfect riding won the hearts of the old school of Mehemet-Ali, which was more adept at mounting a horse than in mental pursuits. On November the 30th, 1854, you were out in the desert with Saïd. The Viceroy's tent was pitched upon an eminence formed

of loose stones.   You had observed that there was a
spot where you could jump with your horse over the
parapet, and this was the route you chose.   You
ought by good right to have broken your neck, but
in the East a rash act often answers as well as a wise
one.   Your hardihood excited universal admiration,
and that same day the charter was signed.   Saïd from
that hour regarded the piercing of the Isthmus as his
own special work, and he brought to bear upon it the
tenacity of an enthusiast and the vanity of a barbarian.
Within a month from that time you started upon
your first exploring of the desert over which you
were in fifteen years to win so decisive a victory.

   "These fifteen years were like a dream, worthy to
be included in the ' Arabian Nights ' or Massoudi's
' Golden Prairies.'   Your ascendancy over that world
so strangely endowed with rough-and-ready energy was
something incredible.   You astonished M. Barthélemy
St. Hilaire, who could follow you at last no longer.
You were, in short, a king, and you enjoyed the advantages of sovereignty, and learnt the great lesson which
it teaches, that of indulgence, pity, pardon, and disdain.   I have seen myself your kingdom in the desert.
When crossing the Ouadi from Zagazig to Ismailia,
you gave me as guide one of your subjects.   He was,
I believe, an ex-brigand whom you had for a time attached to the cause of order.   While explaining to me the
way to handle an old sixteenth-century musket, which
formed part of his armament, he unbosomed himself

to me of his inmost sentiments, which may be summed
up in unbounded admiration for you. You had your
faithful disciples—I was almost going to say your
fanatics—in the camp of those who might be regarded
as your enemies. At Ismaïlia we met an English
lady who was watching very intently the progress of
your workmen to see whether the prophecies of the
Bible were not being confirmed. She took us to see
some tufts of grass and flowers which the infiltrations
of the sweet-water canal had caused to spring up on
the sand. This seemed conclusive to her, for was it
not written in the 35th chapter of Isaiah that, before
the coming of the Messiah, ' the desert shall rejoice
and blossom as the rose ' ! You had some fancy ready
to suit every one's taste, and supplied them all with a
dream after their own heart.

" The word religion is not too strong to express the
enthusiasm which you excited. Your work was for
several years a sort of gospel of redemption, an era of
grace and pardon. The idea of rehabilitation and
moral amnesty always occupy a large place in the
origin of religions. The brigand is grateful to whomsoever comes to preach a jubilee which has the effect
of creating a new departure. You were kind to those
who came and offered their services. You made them
feel that their past would be wiped out, that their
offences would be absolved, and that they would begin
their moral life anew if they were in earnest to help
you pierce the Isthmus. There are so many people

ready to amend their ways if only one will pass the
sponge over some incident in their career. Upon one
occasion, a whole troop of convicts who had escaped
from some prison on the shores of the Adriatic
swooped down upon the Isthmus as upon a land of
promise. The Austrian consul demanded their surrender, but you spun out the negotiations, and in a
few weeks' time the consul was busily employed in
forwarding the money which these worthy fellows
wanted to send home to their poor relations, perhaps
to their victims. The consul thereupon begged you
to keep them, as you had succeeded in turning them
to such excellent account. In a report of one of your
lectures, I remember reading: 'M. de Lesseps stated
that men were trustworthy and not at all evilly disposed when they have enough to live upon. Man
only becomes evil through hunger or fear.' We
should perhaps add: 'or when he is jealous.' You
went on to say: 'I have never had to complain of
my workmen, and yet I have employed pirates and
convicts. Work has made honest men again of them
all; I have never been robbed even of a pockethand-
kerchief. The truth is that our men can be got to do
anything by showing them esteem and by persuading
them that they are engaged upon a work of world-
wide interest.'

" You have thus caused to blossom once more a flower
which seemed faded for ever. You have given in this
sceptical age of ours, a striking proof of the efficacy

of faith and verified in their liberal sense that lofty saying : ' I say unto you that if ye have faith as a grain of mustard-seed, ye shall say unto this moun- tain, remove hence to yonder place, and it shall remove.' The devotion of your staff was immense. I spent a night at Chalouf-el-Terrabah, in a hut inhabited only by one of your employés. That man filled me with admiration : he was convinced that he was fulfilling a mission, he regarded himself as a sentinel placed in an advanced post, as a missionary of France, and an agent of civilisation. All of your men believed that the eyes of the world were fixed upon them and that every one was interested in their doing their duty.

"It is all this, Sir, that in electing you we were anxious to recompence. We are incompetent to appreciate the work of the engineer; the merits of the administrator, the financier, and the diplomatist are not for us to discuss ; but we have been struck by the moral grandeur of the work, by this resurrection of the faith, not the faith in any particular dogma, but the faith in humanity and its brilliant destinies. It is not for the material work which we crown you, for the blue riband which, as we are told, would earn for us the esteem of the inhabitants of the moon, if there were any. No, that is not what constitutes your glory. Your glory consists in having set stirring this latest movement of enthusiasm, this latest manifesta- tion of self-devotion. You have renewed in our time the miracles of ancient days. You possess in the highest degree the secret of all greatness, the art of making yourself beloved. You have succeeded in forming out of incoherent masses a small but compact army, in which the best qualities of the French race have appeared in all their *éclat*. Thousands of men have found in you their conscience, their reason of being, their principle of nobility or of moral renovation.

"The amount of valour, bravery, and resources of every kind which you have expended in this struggle is something prodigious. What a fund of good humour, more especially, must you not have needed to answer patiently the many puerile objections which were raised: the moving sands of the desert, the bottomless mud in Lake Mensaleh, the threats of an universal deluge brought about by the difference in level of the two seas! During the first two years your activity knew no bounds; during that time you travelled twenty-five thousand miles a year, more than the distance round the world. You had to convince Europe, especially England, our great and dear rival. You conformed your habits to those of the country. You went from town to town, with only one companion, taking with you enormous maps, loaded with pamphlets and prospectuses. When you arrived in a town, you went to the mayor or the principal person of the locality, to offer him the chairmanship of the meeting; then you selected your secretary, and after that called upon the editors of the local papers. In that way you held thirty-two meetings in the principal towns of the United Kingdom in forty-five days. You spent your nights in correcting the proofs of your previous day's speeches, and you took away with you a thousand copies, which you distributed the following day.

"You do not scruple to use any of the means which our century has made the essentials of success. You do not disdain the press, and you are right; for, so far as regards its effect upon the public, the manner in which a fact is related is far more important than the fact itself. The press has in our day taken the place of what formerly brought men into communication with one another, viz., correspondence by letter, public speaking, books, and, I might almost add, conversation. To renounce the use of this powerful engine is to renounce one's legitimate share in human action. There are, I am well aware, many Puritanic persons who are content with being right in their own eyes, and who regard it as a humiliating obligation to be right in the eyes of other people. I have an infinite respect for this view, but I am afraid that there is some little historical mistake about it. In former days people gained the good-will of the sovereign and the court by methods very little better than those with which, in our day, the favour of the public is courted. The public at large are guided by their newspaper; Louis XIV. and Louis XV. saw through the narrow spectacles of those about them. Turgot, the most modest of men, had only to convince four persons of his merit: first of all, Abbé Very, his fellow-student in the Sorbonne, a man of very enlightened mind, who spoke of him with great admiration to a very clever woman, Madame de Maurepas; she mentioned him to her husband, and he presented him to Louis XVI. With universal suffrage the candidature is not quite so simple an affair. But there is a reverse to the medal. All that was needed to bring about the fall of the Minister who alone might have saved the monarchy were a few courtiers' epigrams and a change in the views held by Maurepas. What a long chapter might be written anent the blunders of a limited suffrage! Our time is not more frivolous than those which preceded it. We are told that this is the reign of mediocrity. Well, sir, this reign began some time ago. The sum of good sense which emerges from any given society for the purposes of government has always been very small. The man cast in a higher mould who is anxious to do what is right has always been obliged to lend himself to the weakness of the masses. Poor humanity! In order to be of service to it, one must adapt oneself to its measure, speak its language, adopt its prejudices, and enter with it into the workshop, the slums, the lodging-house, and the tavern!

"You did well, therefore, not to allow yourself to be baulked by the petty susceptibilities which, if they were taken too much account of, would make inactivity to seem the highest wisdom. The days are dark; we are working in the night; let us work on nevertheless. The Preacher spoke well when he said that no one can tell whether the inheritor of the fortune which he has built up will be wise or a fool. But did this gifted philosopher draw thence the conclusion that we should do nothing? Not at all. An inward voice urges us on to action. Man does great deeds by instinct, just as the bird wings its flight, guided by a mysterious map which it carries within its tiny brain.

"You have not disguised from yourself the fact that the cutting of the isthmus would serve alternately very varied interests. The great saying, 'I have come to bring not peace but war,' must have frequently recurred to your recollection. The isthmus cut becomes a strait—that is to say, a battle-field. One Bosphorus had sufficed till now to give trouble enough to the world. You have created another, much more important than the first, for it does not place in communication two parts of an inland sea. It serves as a passage of communication between all the great seas of the world. In case of maritime war it would be the supreme interest, the point for the occupation of which the whole globe would make a rush. You have thus fixed the spot for the great battles of the future.

"What more can we do than ring round the field

in which these blind forces meet, than favour in their struggle towards existence all these obscure things which groan, and weep, and suffer before being born? No disappointments shall stop us, we mean to be incorrigible; even amid our disasters works of universal importance still continue to tempt us. The King of Abyssinia has said of you, ' Lesseps is of the tribe of light.' Truly, this king speaks words of truth. We all belong to that tribe. It is a rule in war to march in the direction of firing, from wherever the sound comes. The duty of us civilians is to march towards the light, often without quite knowing whither it is leading us.

"You have rendered such full justice to Henri Martin, your illustrious predecessor, that I need scarcely revert to the subject. He was an excellent citizen, and in all things his thoughts were those of France. When the country took a step forward in that which appears to have been his favourite policy, he followed it; sometimes he even preceded it; but in all things he was sincere. The word of command which he appeared to receive from without in reality came from himself, for he was in perfect harmony with the circle in which he lived. He espoused all the prejudices of which common opinion is composed so honestly that he came to mistake them for primitive and increated truths. But as he was a true Liberal, he experienced no regret when his firmest conclusions were arrested for a stage. He desired that progress should be made by the amelioration of men's intelligence and by persuasion. He may have had his illusions like the rest of us, but he never allowed himself to be blinded except when doubt might seem to him a want of generosity, a sin against faith.*

*   *   *   *   *

"You have been wise, indeed, sir, to place the centre of gravity of your existence above these heart-rending uncertainties of politics, which often leave one only the choice between two blunders. Your glory will not suffer from any intermission. Already you have almost entered upon the enjoyment of the judgment of posterity. Your happy, vigorous, and honoured old age recalls that of Solomon, less, no doubt, its weariness. As to that, you have never known what it meant; and although you have been very well placed to see that all is vanity, I doubt whether that thought has ever suggested itself to you. You must be very happy, sir; satisfied with your life, and indifferent to death, for you are brave. You feel somewhat uneasy, you said in one of your lectures, when you reflect that on the day of judgment the Creator may reproach you for having modified His handiwork. But let me assure you that there is no ground for fear on this score. If there is one person more than another as to whose attitude

---

* Note of the Translator.—I have omitted here the remainder of M. Renan's remarks on the literary career of Henri Martin.

in the Valley of Jehoshaphat I am under no apprehension it is you. You will continue there to
play the charmer's part, and as to the Great Judge,
you will win Him over to you. You have improved His work; He will assuredly be well pleased
with you.

"In the meanwhile, you will come and rest yourself in our company after the indefatigable activity
which you have made the rule of your life. In the
intervals between your voyages from Suez to Panama,
and from Panama to Suez, you will communicate to
us your fresh observations as to the world, whether it
is improving or degenerating, whether it is growing
younger or older; whether, in process of time, as
isthmuses are pierced, the number of lofty and kindly
souls increases or diminishes. Our lives, mostly
passed in the shade, will be supplemented and completed by yours, all of which is spent in the open air.
For my own part, I never see you without fancying
what we might have accomplished together, if we had
associated to found some work in common. And, indeed, if I was not already an old man, I am not sure
that I should not propose some seductive and beneficent scheme to you. But in order to do that, I
should be obliged to resign my post in the Académie
des Inscriptions et des Belles Lettres, the pure and
absolute friend of truth. This I shall never do, for I
derive too much pleasure from my connection with it.
And then the world is so strange; as a rule it will

not allow that a man can be an adept at more than
one thing. The world listens to you when there is
an isthmus to be cut in twain; and there are certain
questions with respect to which it is pleased to give
me a favourable ear. Upon other subjects we are not
consulted, though we might, perhaps, have some good
advice to offer. The will of Providence be done;
we must not complain of the part which has been
assigned to us.

"Yours, assuredly, was a very enviable one. Next
to Lamartine, you have, I think, been the most beloved man of our century—the man upon whom the
greatest number of legends and dreams have been
built. We thank you, as we thank the great poet
who is seated by your side, and who introduces you
into our company, for having afforded—at a period the
great defect of which is the spirit of jealousy and
detraction—to our downcast people the opportunity
of exercising the noblest faculty of the human heart,
that of admiration and love. The nation which knows
how to admire and love is not at the point of death.
To those who tell us that the bosom of this people
has ceased to beat, that it has lost the faculty of
adoration, and that the spectacle of so many abortive
efforts and disappointments has extinguished all its
confidence in what is good, all its belief in what is
great, we reply with the names of you our two
beloved and glorious colleagues. We recall the worship which is paid you, these wreaths, these fêtes

which as a rule are only celebrated after death, and above all, those flutterings in the heart of the multitude which the names of Victor Hugo and Ferdinand de Lesseps ever awaken. This it is which consoles us and bids us say with all confidence, 'Hapless and dear land of France! no, thou wilt not perish, for thou still lovest and art still beloved.' "

THE END.

PRINTED BY J. S. VIRTUE AND CO., LIMITED, CITY ROAD, LONDON.